James Crnkovich's

Atomic America

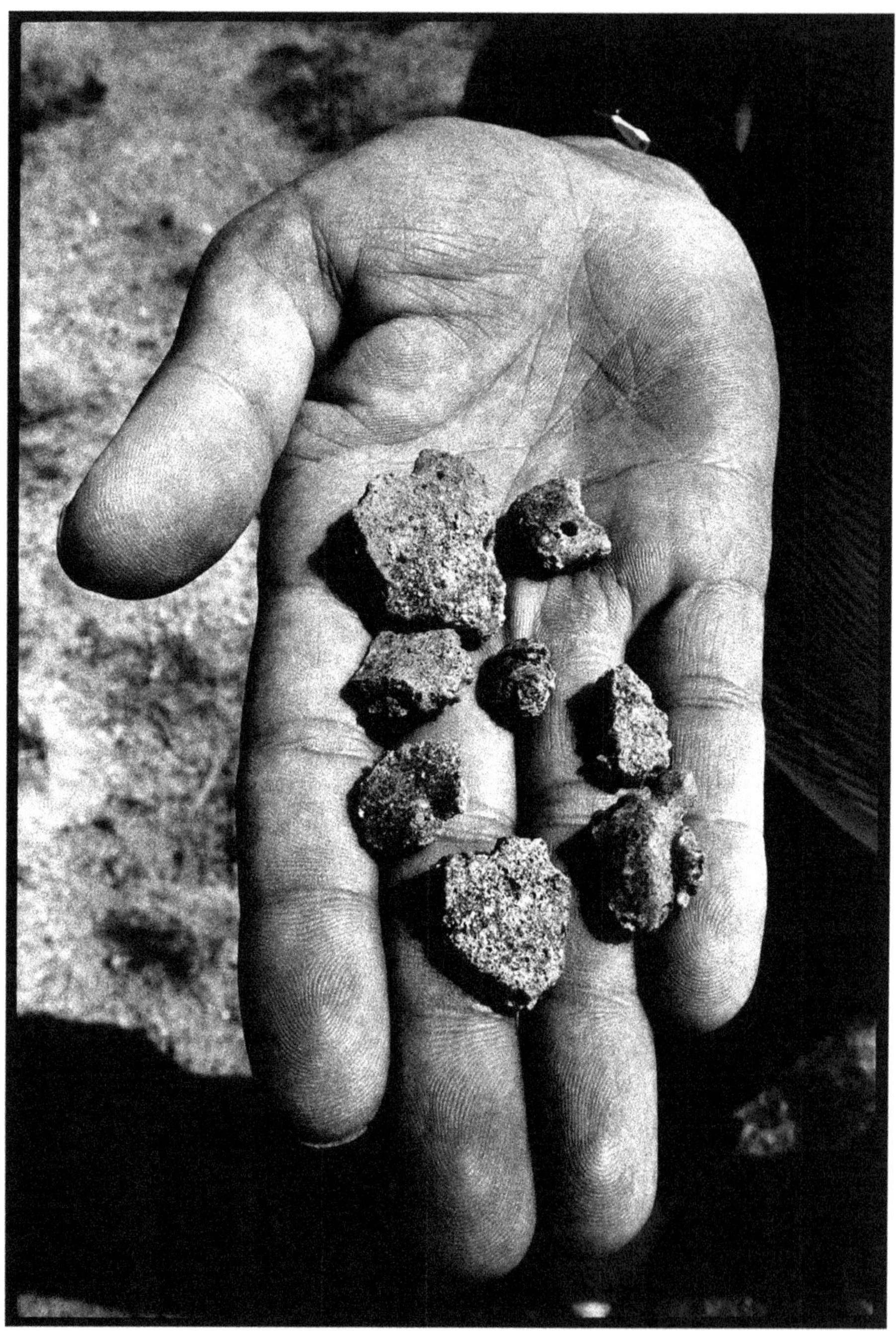

A Handful of "Trinitite." "Trinitite" was created by heat from the first atomic blast that fused desert sand into radioactive green glass on July 16, 1945. This substance is still mildly radioactive today. "Trinity" was the code-name for the world's first atomic explosion. *Jornada del Muerto, Alamogordo Desert, New Mexico, 1989.*

James Crnkovich's

Atomic America

Introduction by Bob Mielke

Afterword on Nuclear Seeing by Robert Del Tredici

Naciketas Press
715 East McPherson
Kirksville, Missouri 63501
2015

ISBN 978-1-936135-08-0 (1-936135-08-6)

Library of Congress Control Number: 2015906146

Published by:
Naciketas Press
715 E. McPherson
Kirksville, Missouri 63501

Available at:
Naciketas Press
715 E. McPherson
Kirksville, Missouri, 63501
Phone: (660) 665-0273
http://www.naciketas-press.com
Email: ndelmoni@gmail.com

An Appreciation

During a long summer trip through the American Southwest back in 1988, my writing partner Bob Mielke looked up from the book *Star Warriors* and said "Listen to this—the writer asks what would happen if an atomic bomb explodes over Milwaukee." The author, it turns out, was Bill Broad, who once was a member of our high school's Media Diggers club, an extra-curricular group of students who explored popular culture by visiting the Milwaukee Art Museum, watching foreign films, shooting super-8 movies, and creating psychedelic backgrounds for bands.

Mielke and I were ardent Media Diggers and worked on all kinds of projects. An Elvis infatuation took us to Memphis more than once. Bob formed a band with his student Steve Davis; they called themselves The Elvis Twins, with Bob playing the decadent and bloated, negative Elvis, Steve being the polite (yes ma'am), generous (handing out Cadillac keys), positive Elvis. Things took a more serious turn when we got inspired to document Minnesota's mineral rich Iron Range through the University of Minnesota's Split Rock Arts Program. Bob Mielke and I immersed ourselves in the lives of Iron Rangers. The project led, by twists and starts, to our most ambitious documentary yet: making visible the all-but invisible US nuclear weapons industry. Fellow Media Digger William J. Broad had gone on to win a Pulitzer Prize in 1986 for reporting on nuclear weapons in the *New York Times*. We felt lucky to have known him back when he was manipulating dyes on a glass tray over an opaque projector and wearing an outfit straight out of Sgt. Pepper. I'd like to think his early high-spirits somehow pushed us, with élan, into this beast of a project. Mielke and I persisted with nuclear people, places, and things, on and off, for thirty years.

We were both still novices looking for a roadmap through the labyrinth of the American bomb when we came upon Robert Del Tredici's seminal book *At Work in the Fields of the Bomb* (1987). His groundbreaking photographs and text laid the foundation for our own long day's journey into the nuclear realm.

Then I met Minneapolis photographer Paul Shambroom. His portraits of weapons and weaponeers inspired me to get closer, push harder, go deeper. At the opening of Shambroom's solo show at the Walker Art Center, Paul introduced me to Robert Del Tredici. We hit it off. He asked me to send photocopies of my work. This opened a floodgate of correspondence. I soon became proud owner of signed copies of his books on The Bomb, Three Mile Island, and Moby Dick, along with over a dozen postcards from Europe, Asia, and the USA. In the summer of 2007 Bob Del Tredici and I, with our two young sons, made a memorable road trip through the nuclear Southwest.

At one point that same summer the two Bob's and I were pondering a batch of my pictures on Mielke's dining room table. Mielke asked Del Tredici if traveling with me was a little like Paul McCartney traveling with a groupie. Without missing a beat Del Tredici quipped, "Which one of us is Paul McCartney?" With that remark Del Tredici told me he considered me his equal. I was elated.

Eight months ago I made another road trip—from Phoenix to Montreal. My trunk held a big box of prints. I needed to ask Bob's help to organize my images into a book. He was just finishing a book on Tibetan guru Chogyam Trungpa and teaching the history of animated film, but he put my project on the front burner. In less than a week we had reduced my 357 prints to 71. It was like distilling down an inchoate epic poem into a series of haikus. At first, Del Tredici's spreads seemed disconnected; but as I looked them over again, and yet again, and as the captions emerged, the whole thing fell into place. To see my own work praised by an artist whose vision I'd admired for years confirmed that I'd been on the right track all along.

My gratitude goes out to both these Bobs. Without the two of them, the document you hold in your hand would not exist.

Remnants of an Averted Inferno
Bob Mielke

In the summer of 1987, I began a collaboration with photographer James Crnkovich to document the Bomb. Nuclear weapons were in my family, though I didn't realize it at the time: my father, it turns out, had worked for the Manhattan Project; and my brother had been a flight surgeon at Area 51 on the Nevada Test Site. And James's father was a high school physics teacher who had spent time in Oak Ridge, Tennessee, where uranium for the Hiroshima bomb was enriched.

We wanted to capture the culture of American nuclear weapons from as many angles as possible, including Bomb workers and their families, protestors and true believers, down-winders from Utah and Nevada, and atomic veterans who'd been marched out over Ground Zero. We also wanted to photograph nuclear weapons hardware, and we saw a lot of it in atomic museums, at memorial sites, and across the pock-marked Nevada Test Site. And then there was all that "atomicana" —my name for the kinky products and nutty signs celebrating the mighty atom in atomic-themed cafes, laundromats, and grocery stores, on candy-wrappers, in cocktails, and dominating all kinds of Bomb memorabilia.

We applied for a grant from the Twin Cities. We didn't get it; but we decided to take on the project anyway and ended up driving around the USA for the space of a decade and beyond, with work still happening intermittently right up to the present day.

Although the Bomb's charisma is most brilliantly delivered in the "money shot" of its mushrooming self-immolation, it was of course no longer possible for us to capture that phenomenon. James did the best he could with the leftovers: these included vast nuclear landscapes and seedy still lifes in abundance. James was also drawn to massive installations, mostly abandoned, built to track incoming enemy missiles. He also captured melted metal shelters from the seared floor of the Nevada Test Site. In Los Alamos he photographed the banal eerie contents of an atomic junk yard called The Black Hole, run by former Los Alamos worker Ed Grothus. I did most of the research on these sites; James's forte was spicing up what we were seeing with visual touches of humor mingled with dread.

Some places we got to tour but could not photograph. One of them was the cavernous Cheyenne Mountain Complex in Colorado, home to NORAD, watcher of the skies; another was the sprawling 300-square-mile Savannah River Site with its five plutonium production reactors. I have, on occasion, known James to discreetly bend the law to get a shot, like the time he trespassed in the company of protesters, who poured blood on a missile silo near

Knob Noster, Missouri.

We made some discoveries in our odyssey through Atomic America. Virtually all the manifestations of atomic culture we came upon confirmed a key idea put forth by Paul Boyer in *By the Bomb's Early Light: American Thought and Culture at the Dawn of the Atomic Age*—namely, that every idea ever held about the Bomb had been expressed within five years of its triple debut at Alamogordo, Hiroshima, and Nagasaki: namely that the Bomb was our savior from war, proof of American superiority, a harbinger of the apocalypse, a transcendent spiritual force, a Darwinian test that only the fittest would survive, a sword that could be turned into a plowshare. It was even sexy (the 1946 bikini swimsuit was named after the first post-Nagasaki atomic blast that ripped through the Pacific Atoll of Bikini).

We saw Boyer's thesis illustrated everywhere we went. It was increasingly evident that we were dealing with ideological strategies to contain a deep ambivalence about this new weapon. Critic Roland Barthes has called this "myth-making"—by which he means the attempt to turn history into nature to make the status quo seem inevitable. The authors of *Nukespeak* reveal how one of the greatest tools for this type of containment and obfuscation is language.

The language of nuclear containment appeared even before the world's first atomic blast at Trinity when Los Alamos weaponeers referred to their as yet unexploded new weapon as "the gadget."

As early as July 5, 1945, President Truman described Hiroshima as "a purely military" target when authorizing the bombing of Hiroshima; but on August 10 he told Henry Wallace he would not authorize a third bombing because he didn't like killing "all those kids" (Boyer, *Fallout* 23). That was one steep learning curve in just over a month.

During our decade on this project, we knew we were in for the long haul and that our ultimate audience would be future generations. Plutonium, after all, has a half-life of 24,000 years. If nuclear weapons were to be banished from the world's arsenals today there would still remain the dark explosive matter of the Bomb, as well as long-lasting environmental consequences of its mass-production. James Crnkovich has captured how it all looked at the turn of the century for eyes as yet unborn to contemplate with all the complex array of emotions that these uncanny artifacts tend to inspire.

Works Cited

Boyer, Paul. *By the Bomb's Early Light: American Thought and Culture at the Dawn of the Atomic Age*. New York: Pantheon, 1985.

—. *Fallout: A Historian Reflects on America's Half-Century Encounter with Nuclear Weapons*. Columbus: Ohio State University Press, 1998.

Hillgartner, Stephen and Richard C. Bell and Rory O'Connor. *Nukespeak: The Selling of Nuclear Technology in America*. New York: Penguin, 1983.

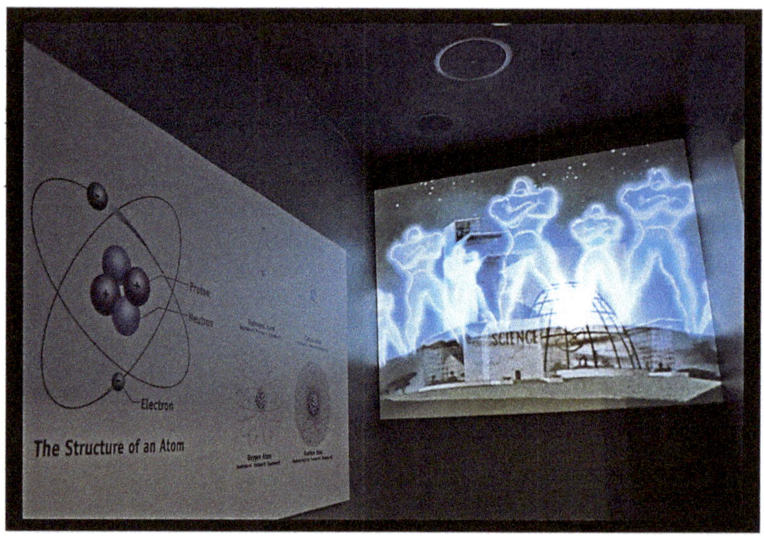

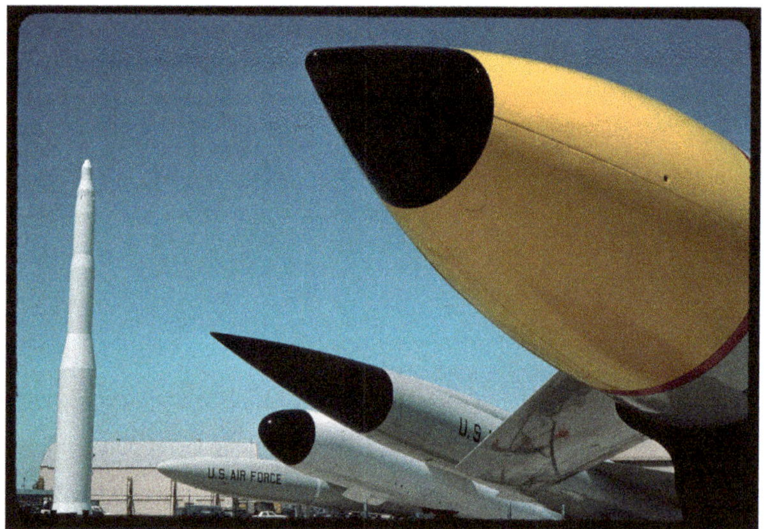

The Age of the Atom is the title of a film from the 1960s that likens the power of the atom to a gang of genies ready to obey all commands of their masters. The film is regularly screened at the *National Atomic Testing Museum, Las Vegas, Nevada, 2011.*

Nuclear Missiles on display behind *Albuquerque's National Atomic Museum, Albuquerque, New Mexico, 1992.*

I consider myself something of a desperado. I remember back in the seventh grade after a big storm I rode a picnic table upside down on a swollen creek through the edge of town. When I was a college freshman my cousin was getting married in Florida so I hopped a freight a thousand miles to the church. I arrived in the middle of the night, sensed I was in a dangerous part of town and shimmied up the cross to sleep on the church roof. A security guard later told me he almost shot me. My adventurous childhood was prelude to the risky business of stalking the Bomb.

During WWII my father was in an officers' training program in Philadelphia. Overnight he watched many of his classmates get shipped to the front lines to become cannon fodder in the Battle of the Bulge. Experiencing the cost of war up close had a big impact on him. When I began my nuclear project, he was quick to point out that as ghastly as the human toll had been in Hiroshima and Nagasaki, if we hadn't dropped the A-bombs, Hirohito would have kept fighting and the U.S. would have suffered steep losses.

My father became a high school physics teacher and studied nuclear physics at Oak Ridge. I was in the sixth grade when he took me to see this sprawling bomb complex. I still remember my fascination with the guard towers we passed as we drove into the city. That same fascination remains with me today.

I admit it: I have a love/hate relationship with the Bomb. I would even say that I have a fetish for all things nuclear—the authentic thing as well as its signs and symbols, residues and relics, its markers and its mighty wreckage. Having acknowledged my own atomic fascinations, I can see the same in others—so I have also aimed my lens at the Bomb's perpetrators and victims, its managers and marketeers, its fellow travelers and ordinary folks just plain curious about humanity's most mega maniacal invention ever.

James Crnkovich

James Crnkovich's Atomic America

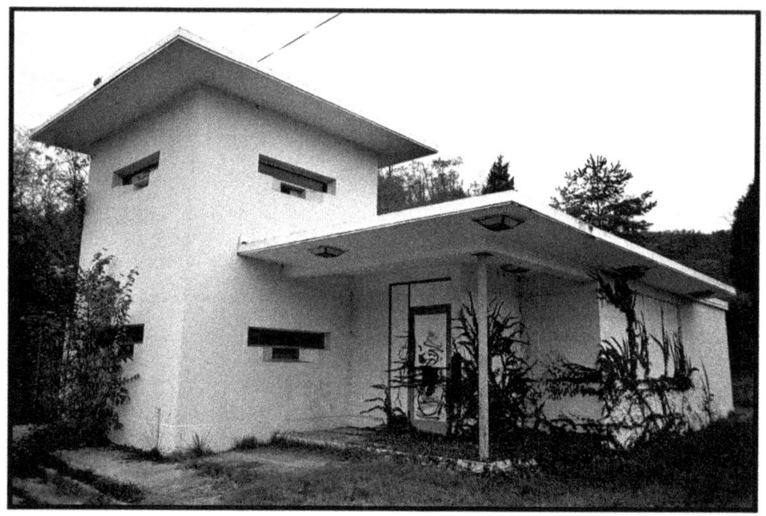

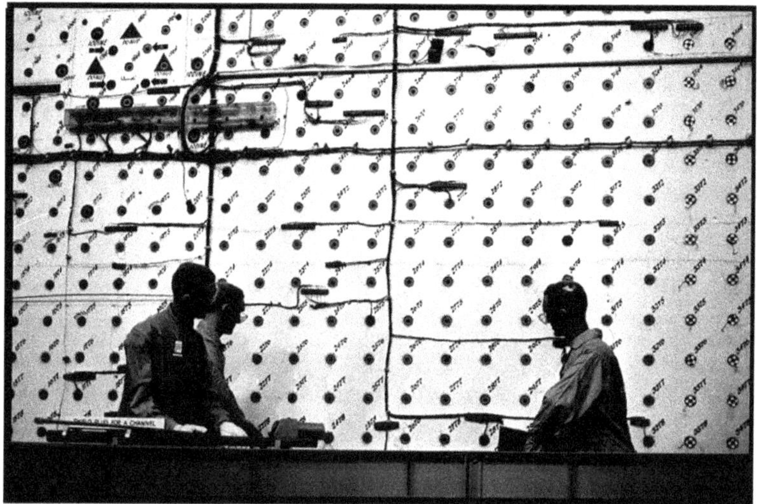

Guard Tower from the 1940s manned with machine guns at the once top-secret *Oak Ridge National Laboratory, Oak Ridge, Tennessee, 1989.*

The X-10 Plutonium Production Reactor provided Los Alamos with its first significant supplies of bomb-grade Plutonium-239 during the Manhattan Project. Nuclear workers used metal rods to control the fission reaction inside the block of graphite forming the core of the X-10 reactor. *Oak Ridge National Laboratory, Oak Ridge, Tennessee, 1989.*

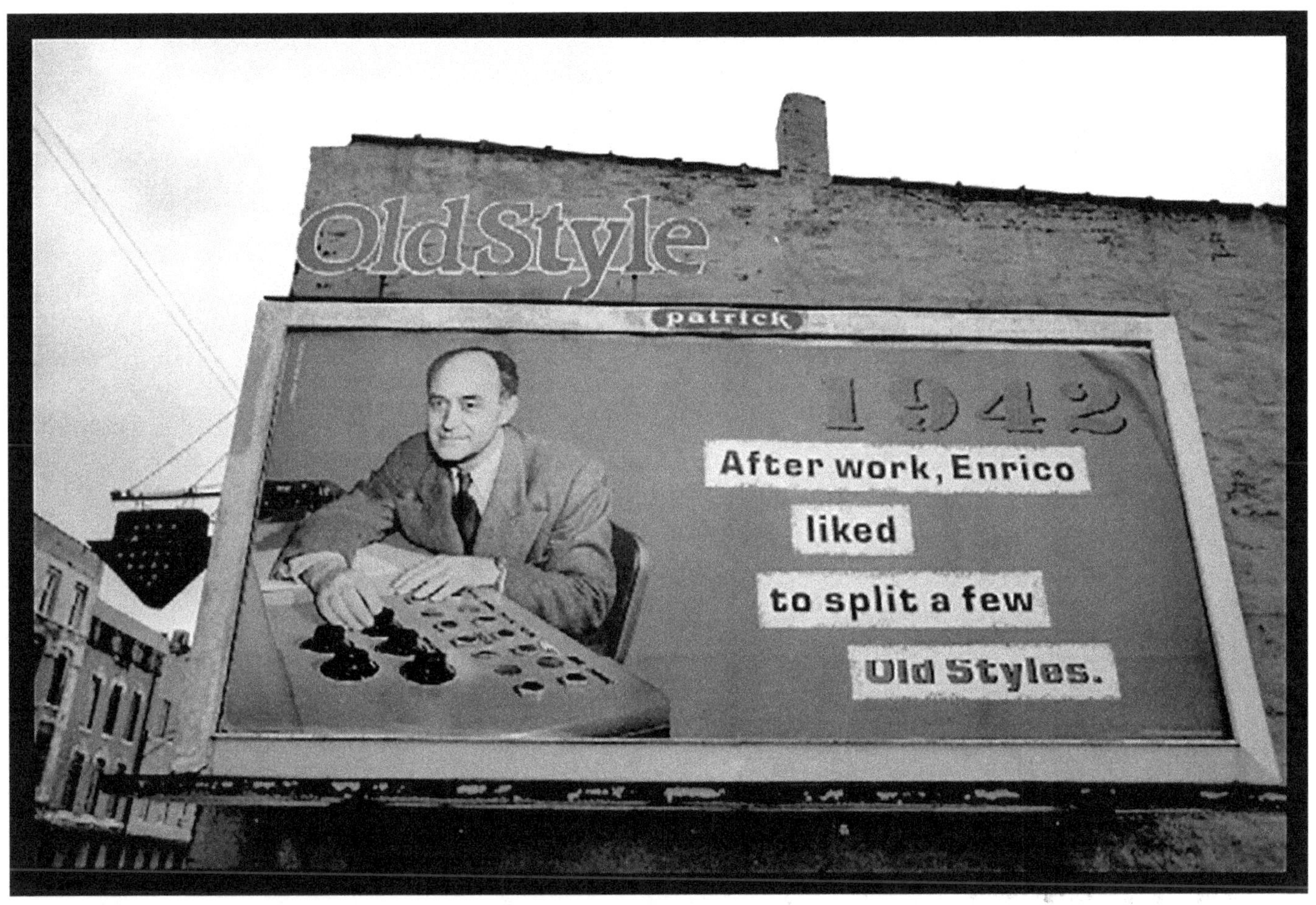

Enrico Fermi pioneered the world's first sustained nuclear fission reaction. A Chicago brewery enjoys wordplay between the master splitter of atoms and the ordinary Joe who liked splitting beers. *Chicago, Illinois, 1994.*

The Nuclear Age is the title of Henry Moore's 12-foot bronze sculpture on the campus of the University of in a graphite pile underneath the Chicago university's squash courts. The sculpture was unveiled on the Chicago

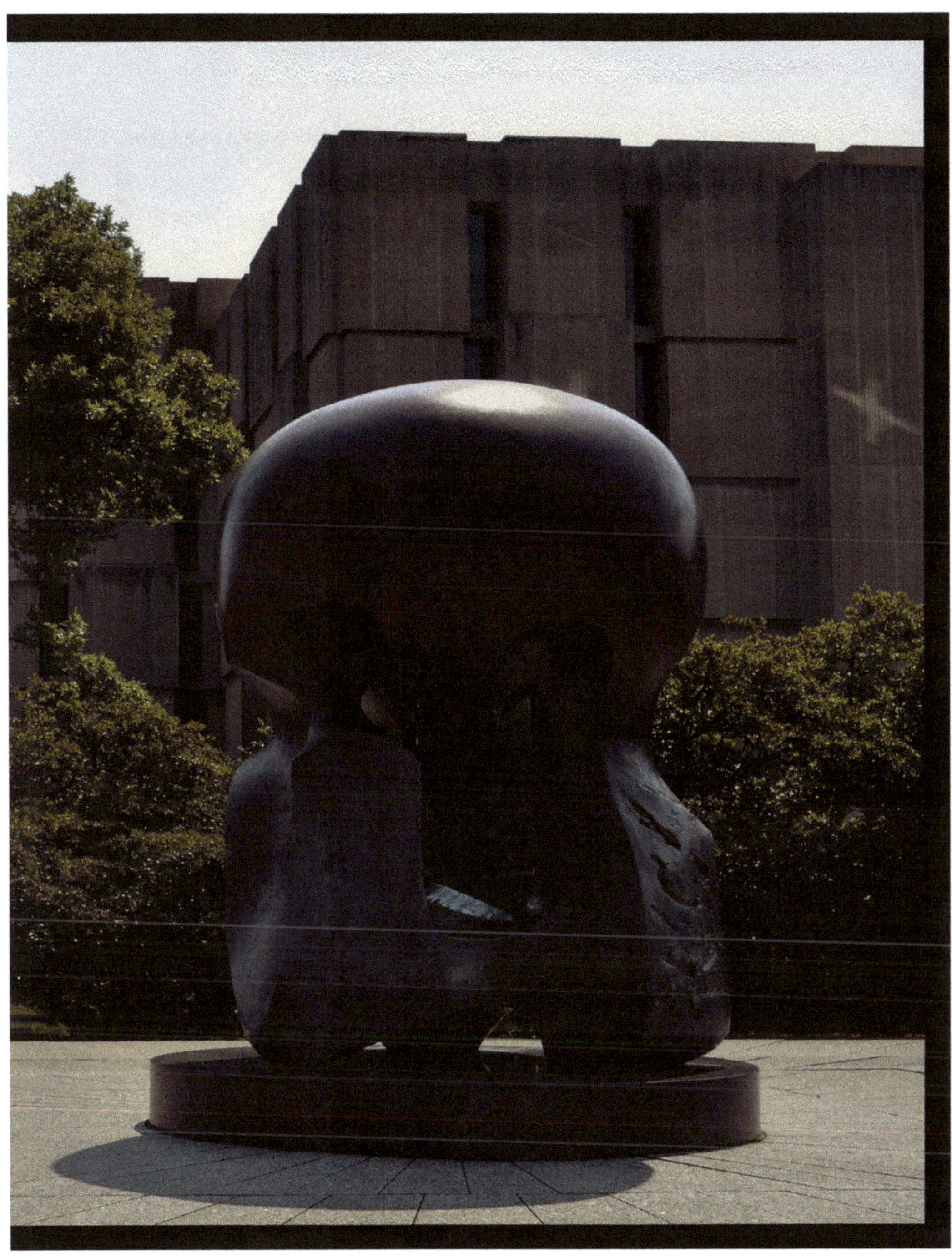

Chicago. It commemorates the world's first sustained nuclear fission reaction which Enrico Fermi engineered campus in 1967. *Chicago, Illinois, 2013.*

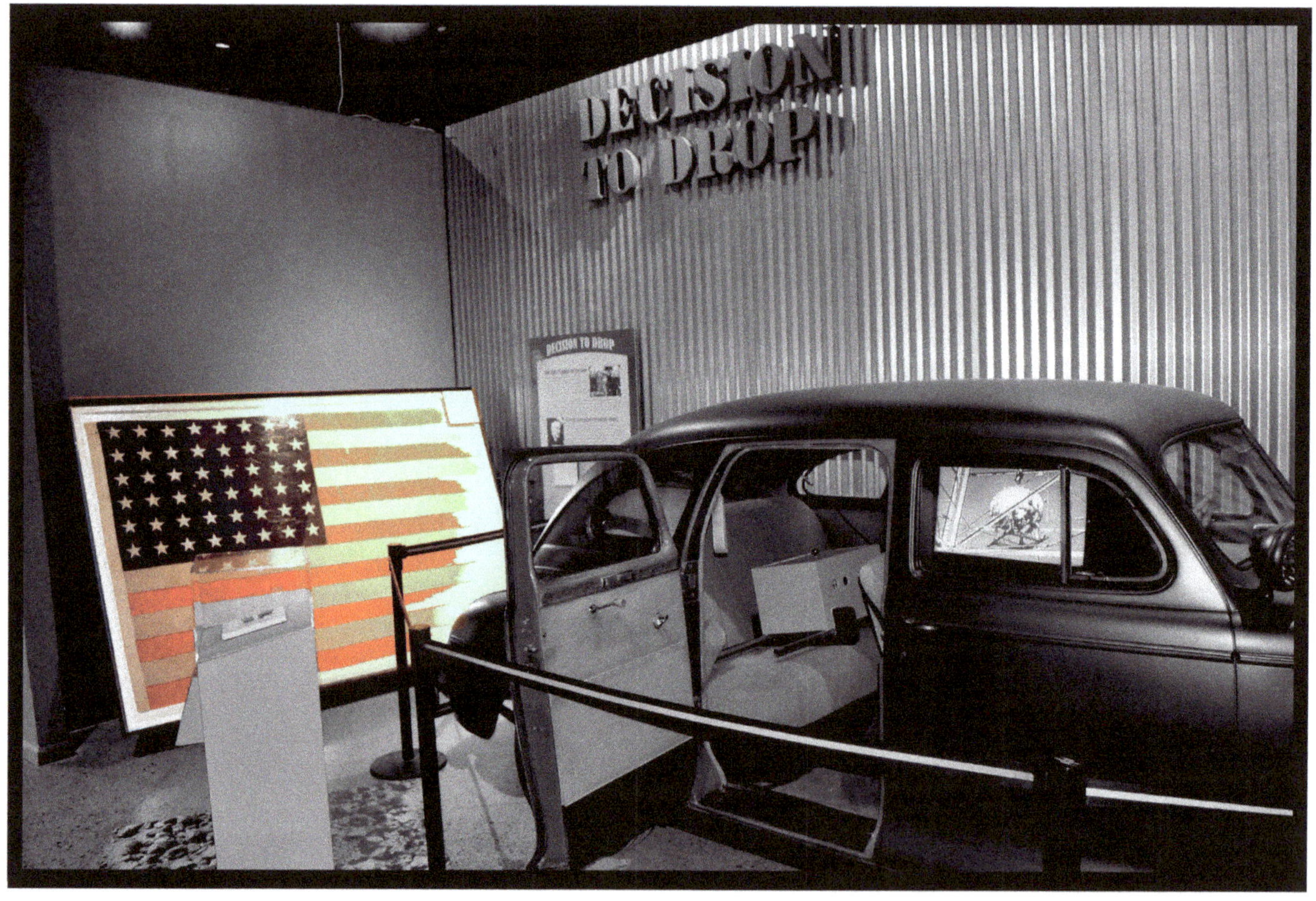

Plutonium Transport Vehicle. In 1945, this brown Plymouth sedan delivered the plutonium core of the world's first atomic bomb to the Trinity Tower in the Alamogordo Desert. *National Museum of Nuclear Science and History, Albuquerque, New Mexico, 2013.*

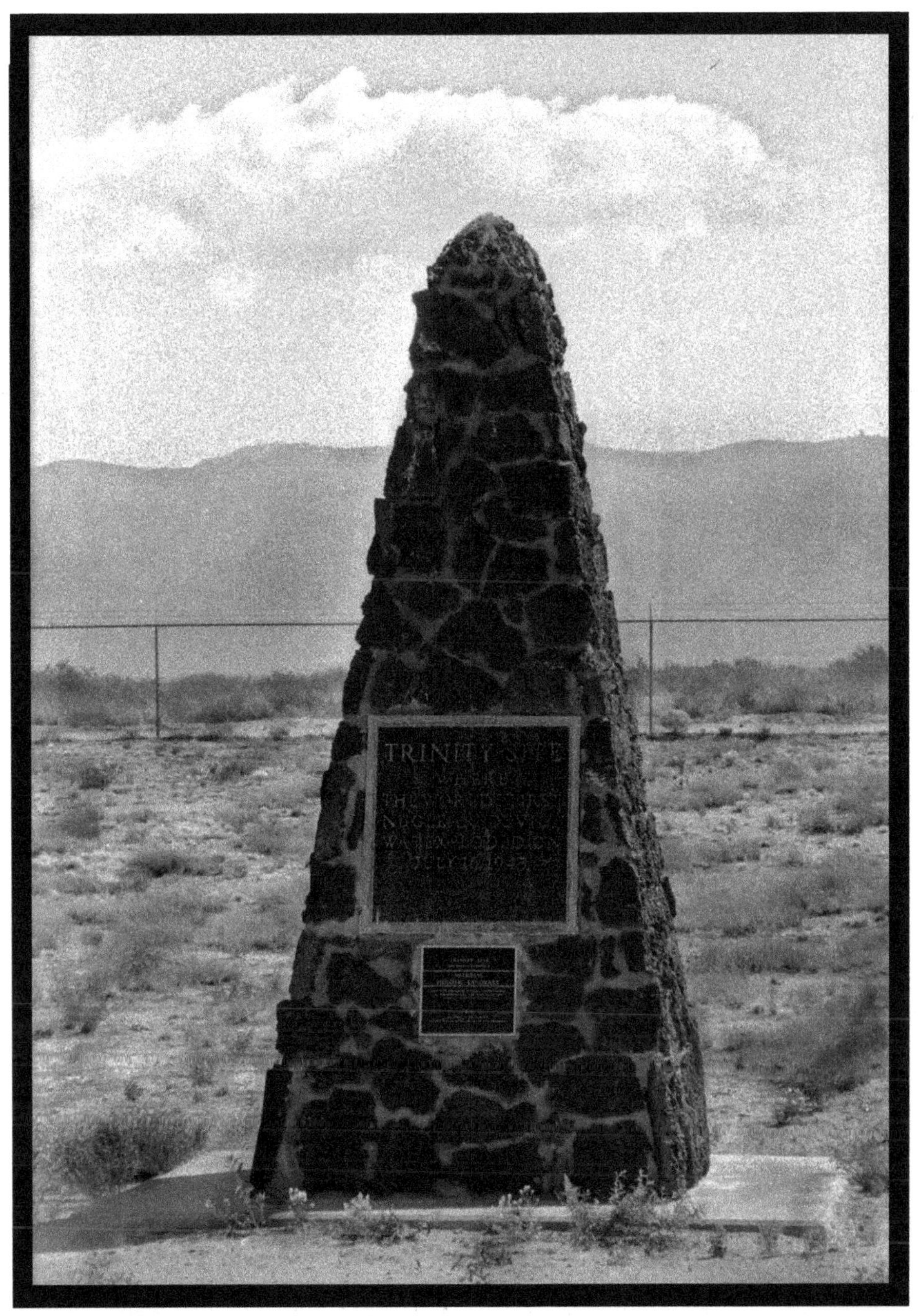

Ground Zero Numero Uno. This modest stone pyramid marks the site of the world's first atomic explosion. It occurred 100 feet above the desert floor on a tower at this spot on July 16, 1945. *Trinity Marker, Jornada del Muerto, Alamogordo, New Mexico, 1989.*

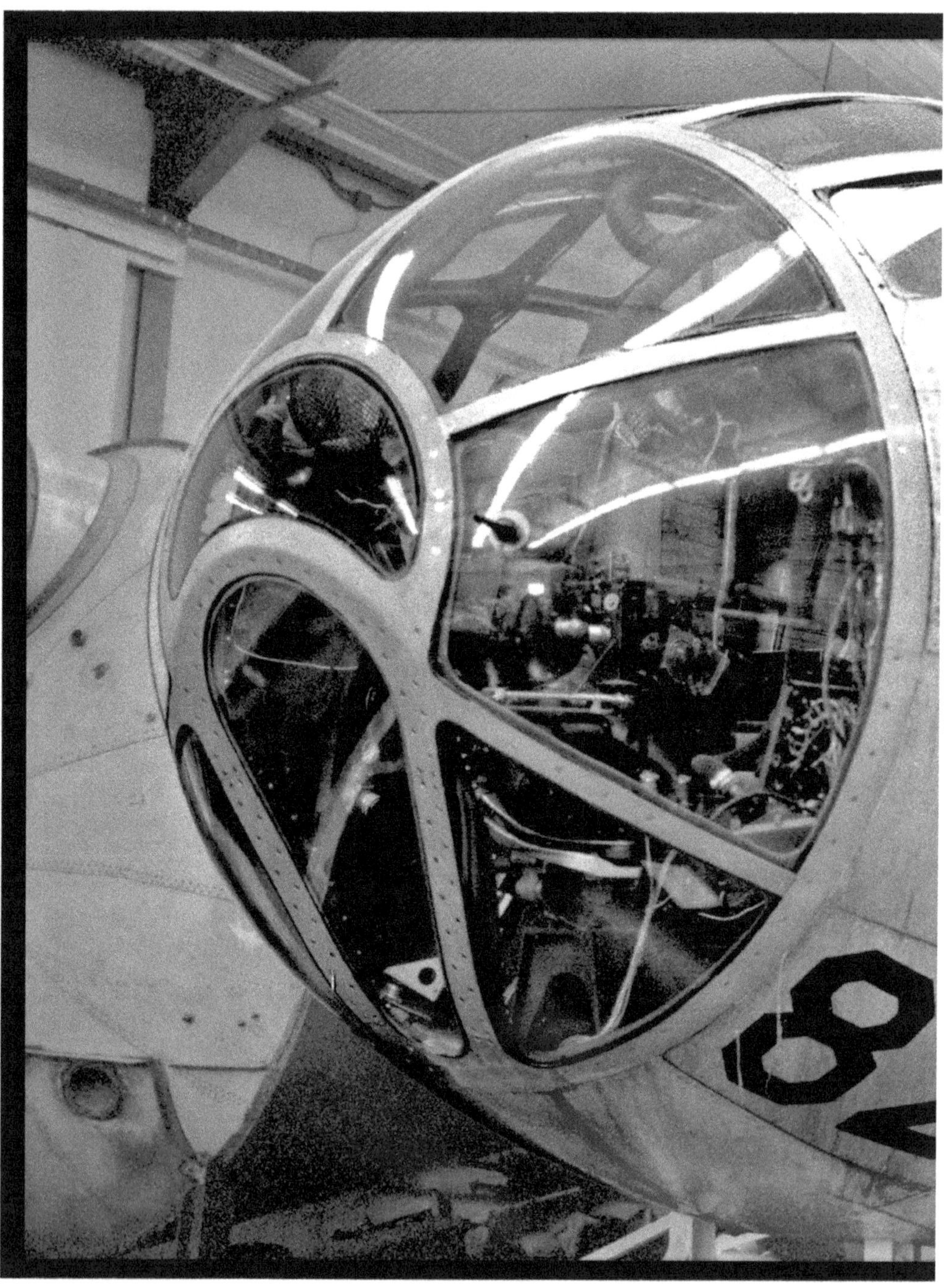

Hiroshima B-29. Paul Tibbets commanded the 509th Airborne Division that dropped the first atomic
The Enola Gay B-29 Superfortress Bomber, Paul E. Garber Facility, Suitland, Maryland, 1990.

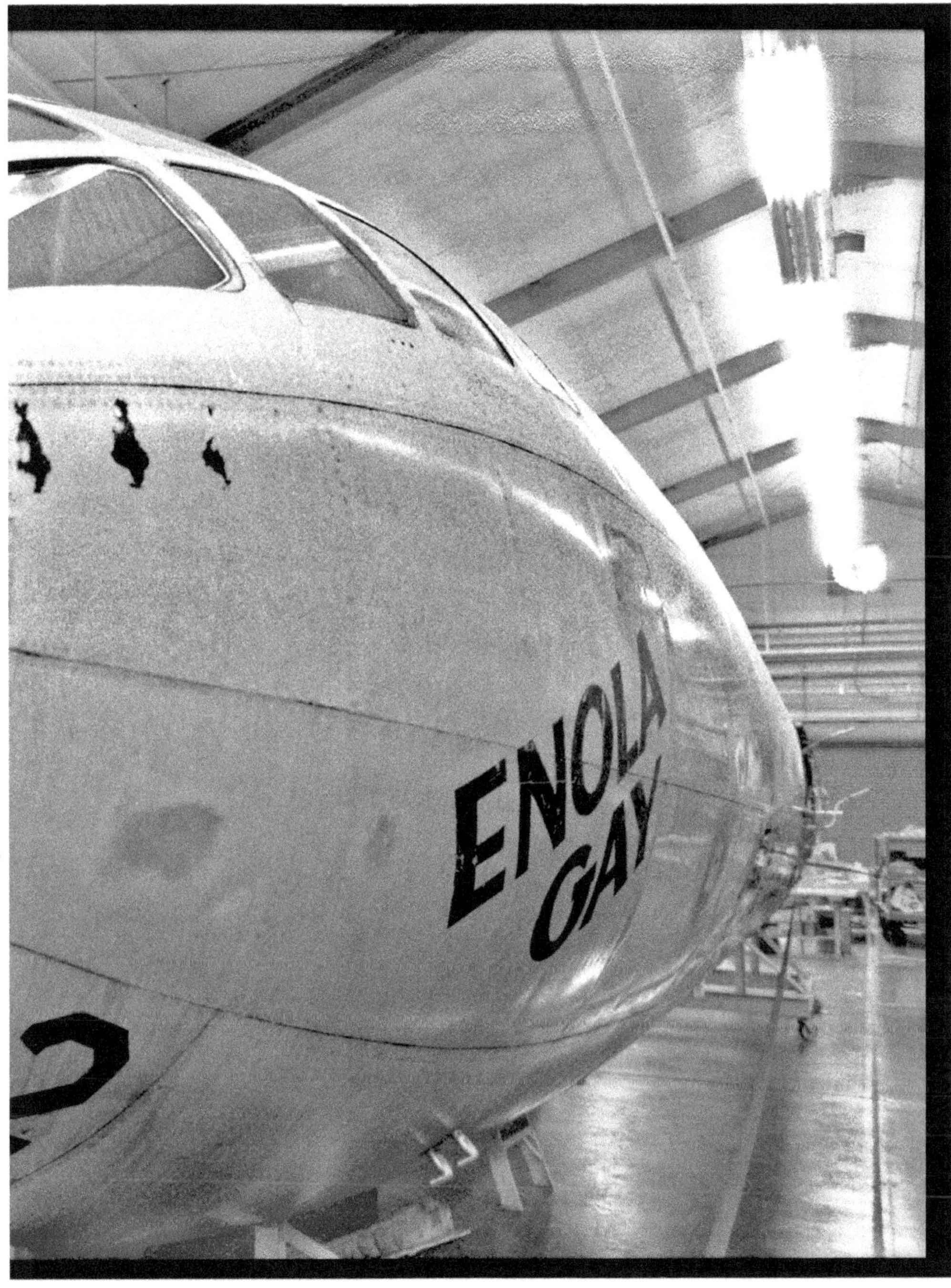

bombs. He named the B-29 Superfortress that a-bombed Hiroshima after his own mother.

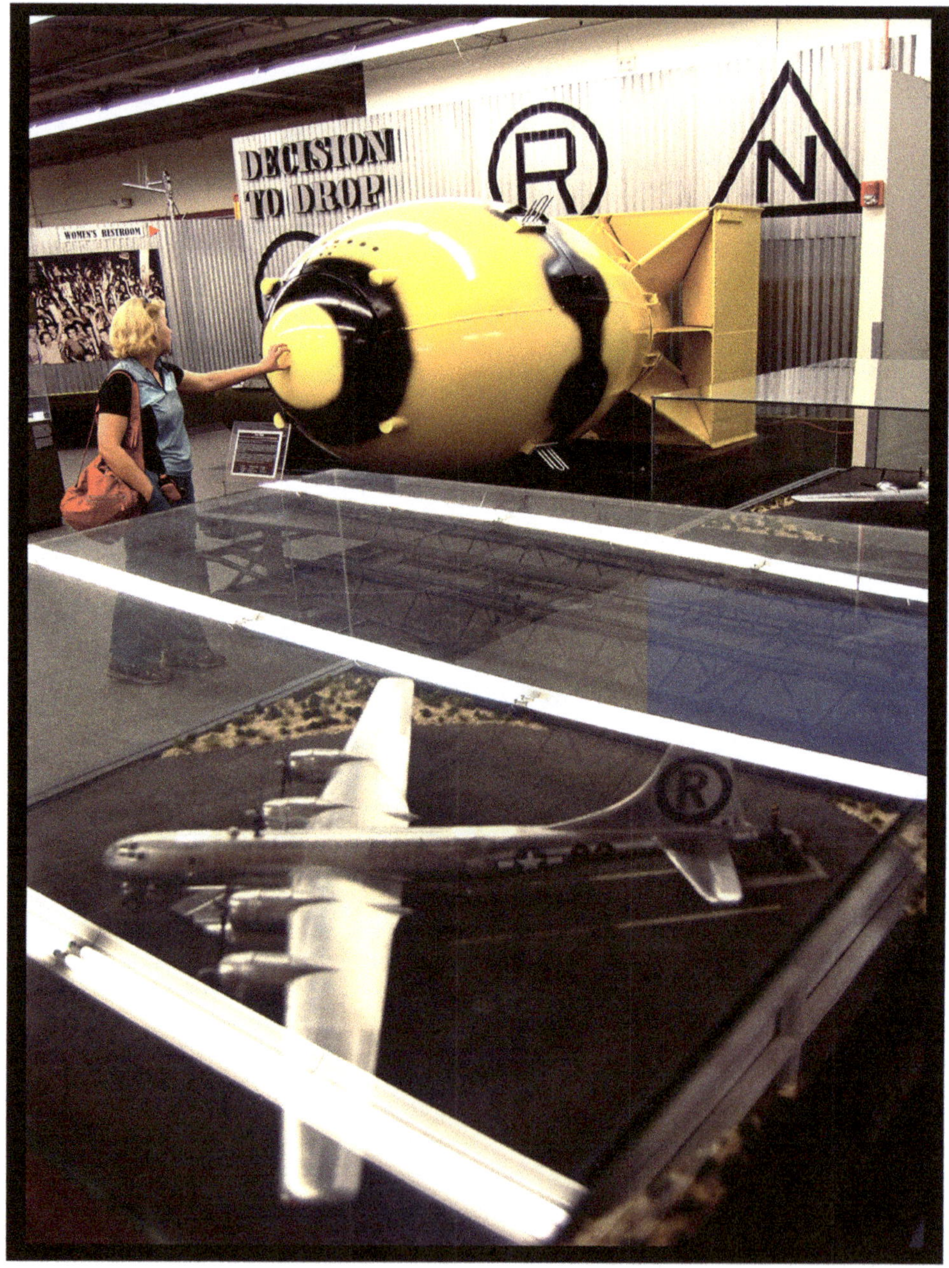

Reaching Out to Touch the "Fat Man," the nickname of the Nagasaki a-bomb. Its core was a sphere of plutonium-239. Plutonium is the nuclear explosive of choice in the American nuclear arsenal—and worldwide. The Fat Man displayed here is a replica of the original outer casing. In the foreground is a model of "Bockscar," the B-29 that dropped Fat Man on Nagasaki. *National Atomic Museum, Albuquerque, New Mexico, 2008.*

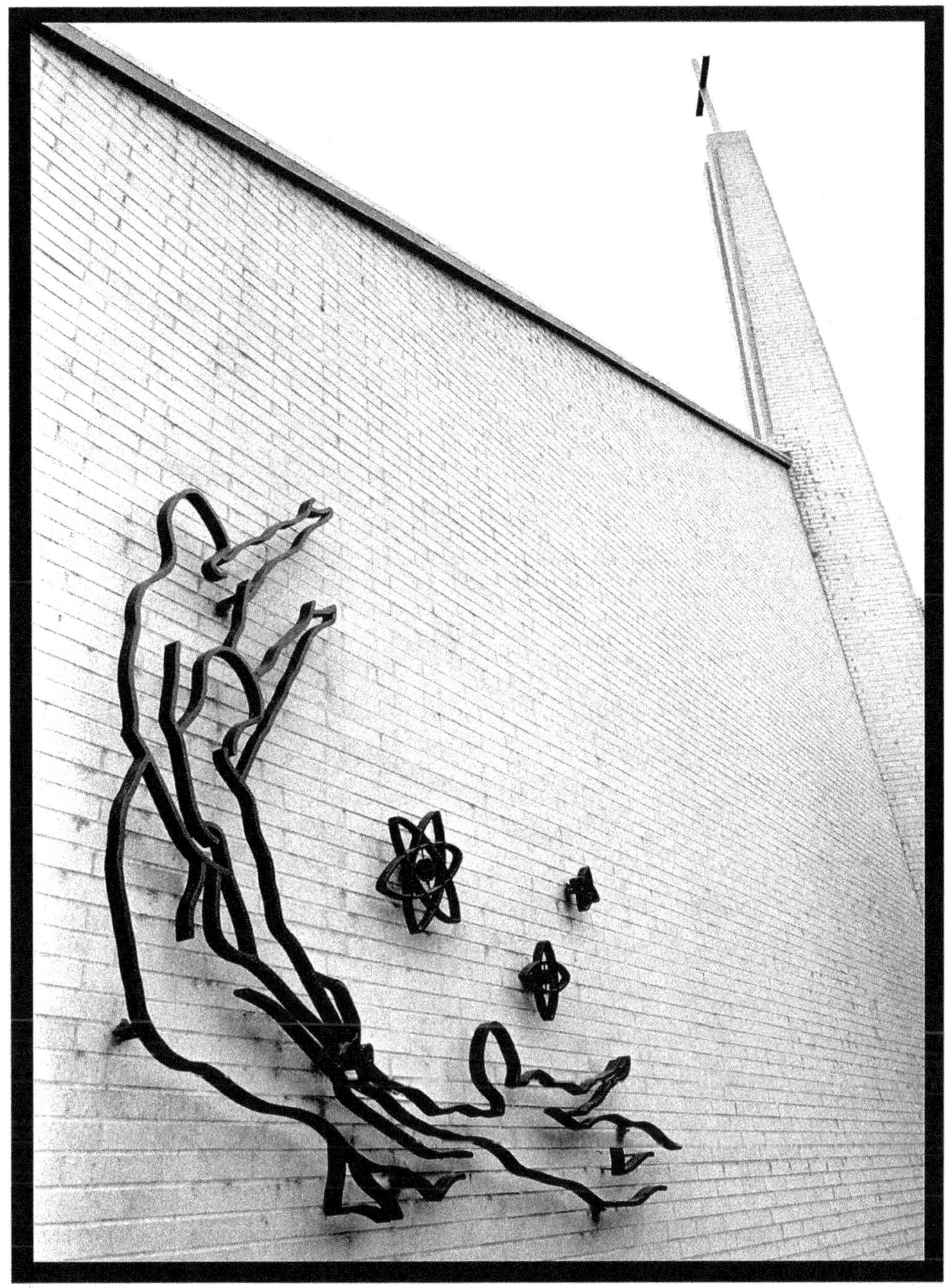

Nuclear Family. Faith in the Risen Lord allows devout mortals to ascend to their heavenly reward through a field of radioactivity. This metallic graphic is affixed to a church ten miles from the *Hanford Nuclear Reservation, Richland, Washington, 1992.*

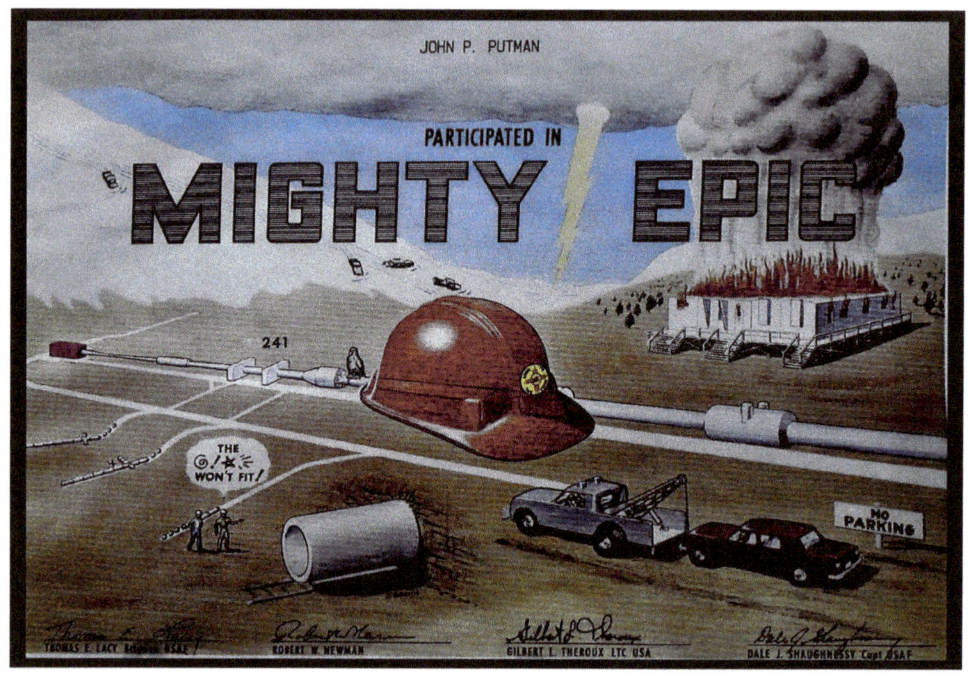

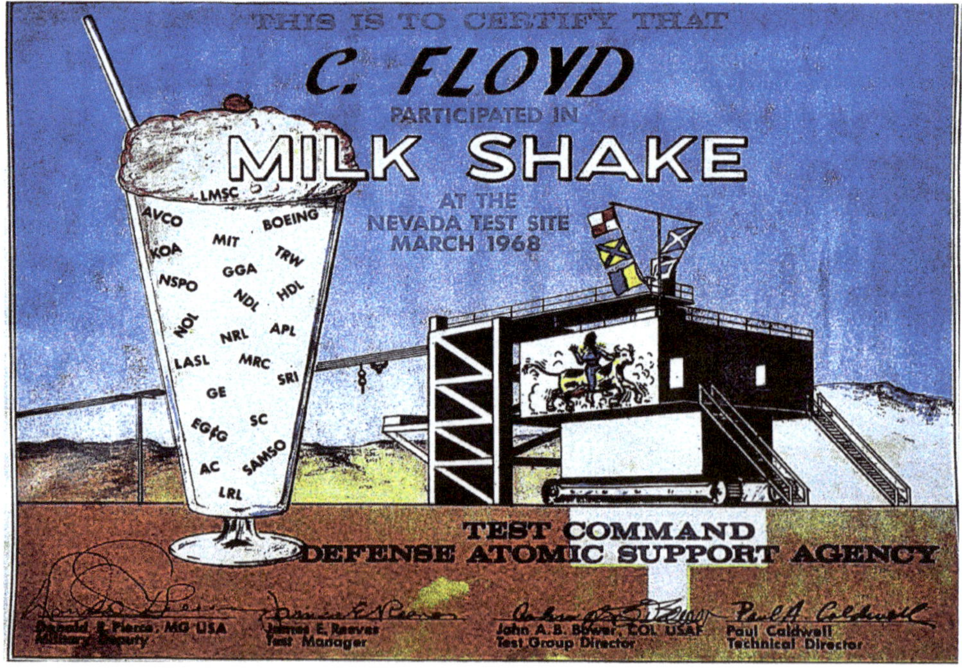

Official Participation Certificates presented to individuals working on specific nuclear bomb tests

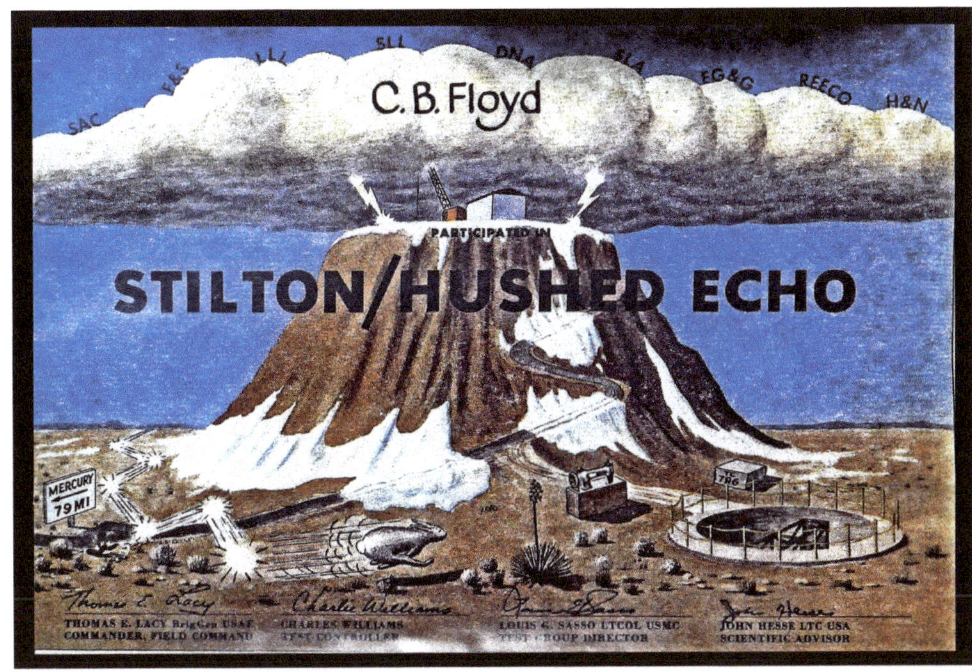

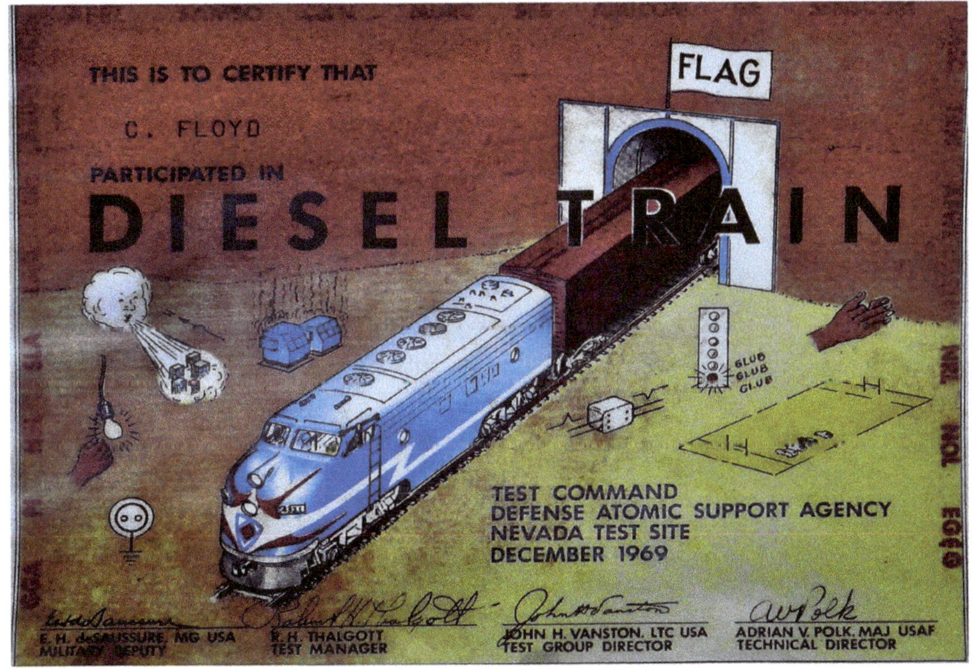

at the Nevada Test Site. *Atomic Testing Museum, Las Vegas, Nevada, 2011.*

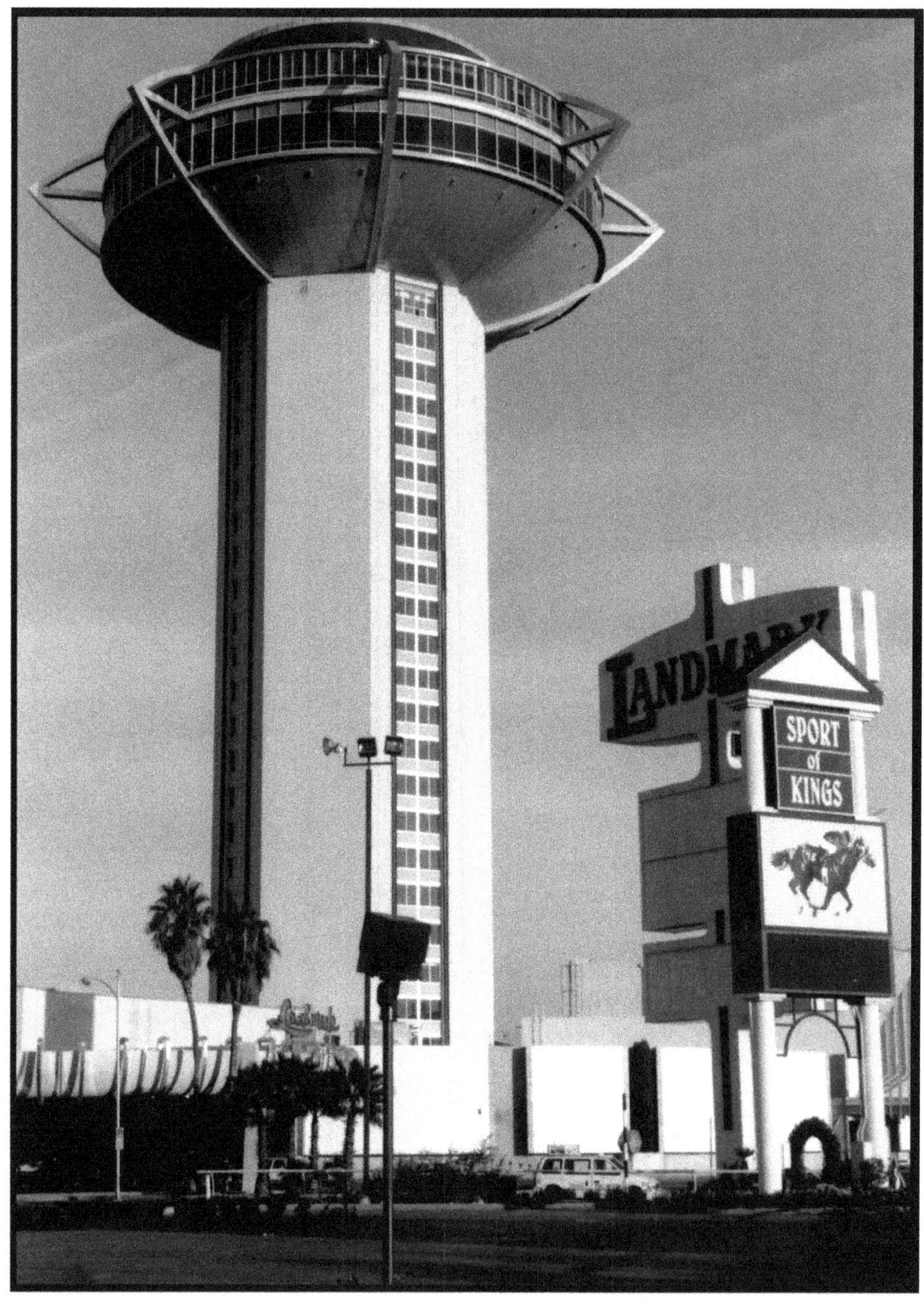

Mushroom Cloud Hotel on the Las Vegas Strip. The structure of the Landmark Hotel mimics a mushroom cloud recalling the nuclear testing program at the nearby Nevada Test Site. *Las Vegas, Nevada, 1989.*

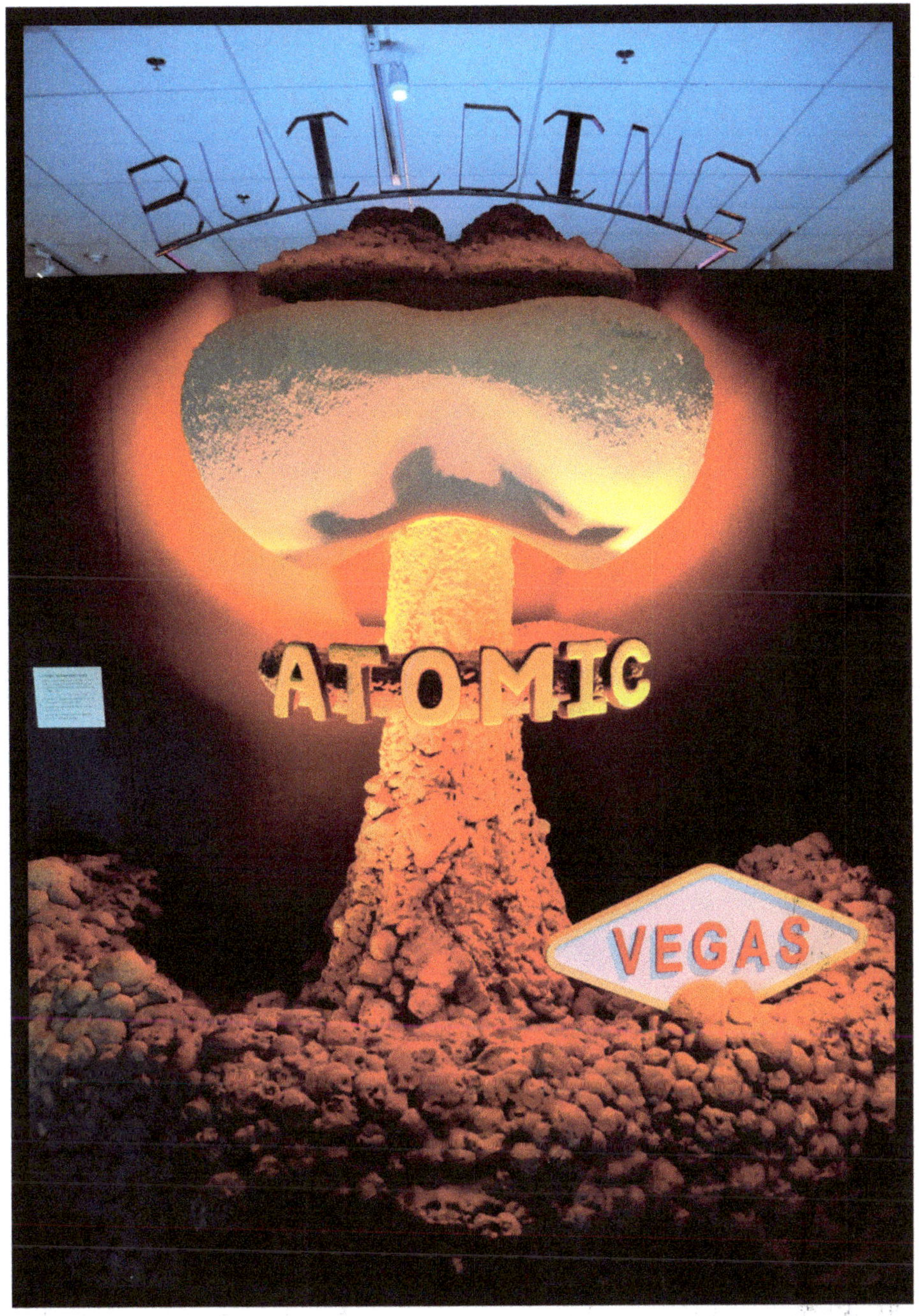

Entrance to the Exhibit "Building Atomic Vegas." As Las Vegas grew, designs inspired by the a-bomb proliferated in the city's culture. It was not unusual for a family member to work at the Nevada Test Site while another member of the same family worked in a Vegas casino. *Atomic Testing Museum, Las Vegas, Nevada, 2011.*

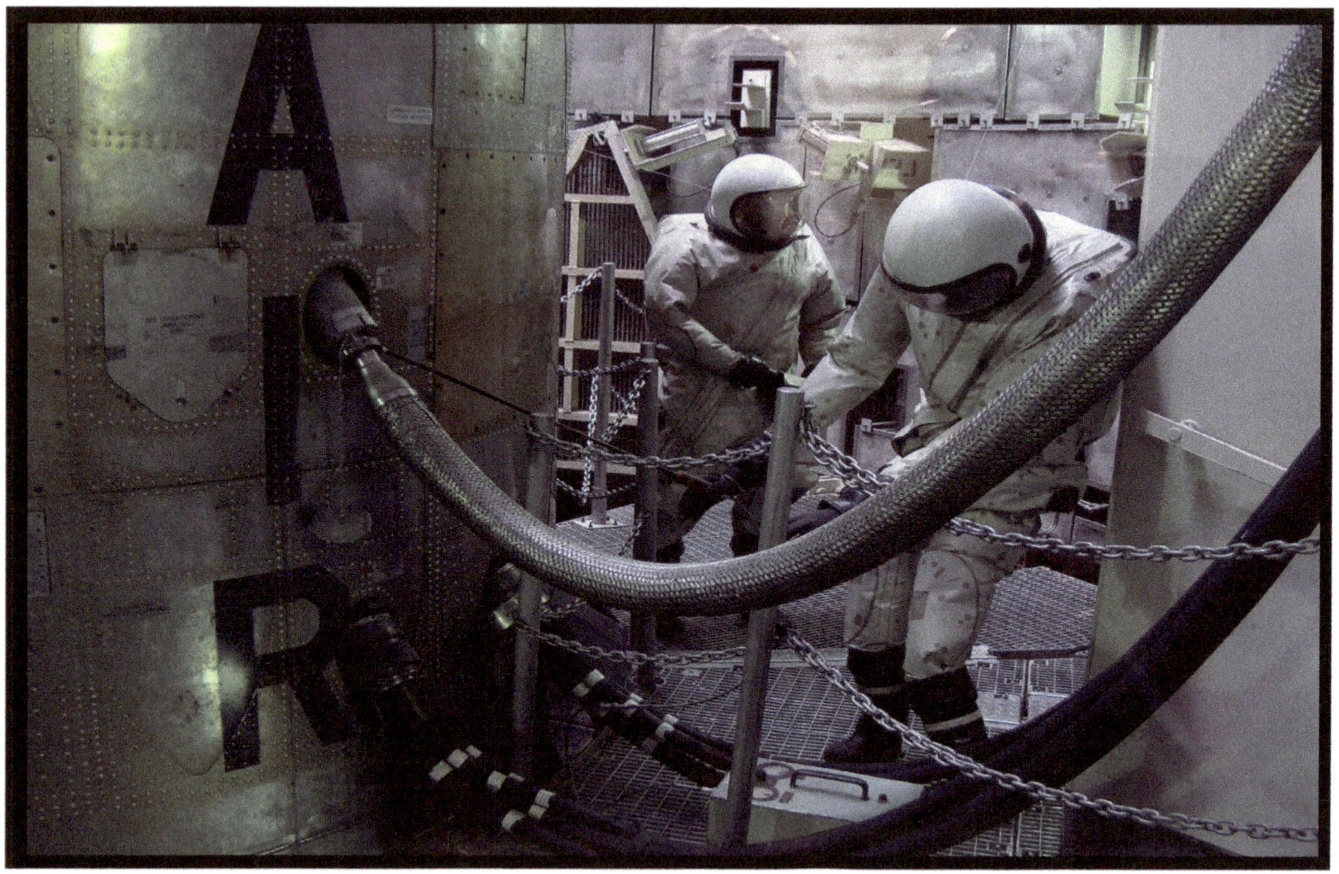

Fueling Titan to Launch Within a Minute. Storing fuel inside the Titan missile silo allowed for faster launches—but at greater risk. Unstable Titan fuel was responsible for the death of three airmen in two fuel accidents: two men were killed in Rock, Kansas, in 1978; in 1980 one man was killed in Damascus, Arkansas, in a fuel fire that destroyed two sites. And in 1965 fifty-three persons lost their lives in a diesel fuel fire that broke out during Titan silo modifications in Searcy, Arkansas. The Titan became the last missile to use liquid fuel. *Titan Missile Museum, Green Valley, Arizona, 2004.*

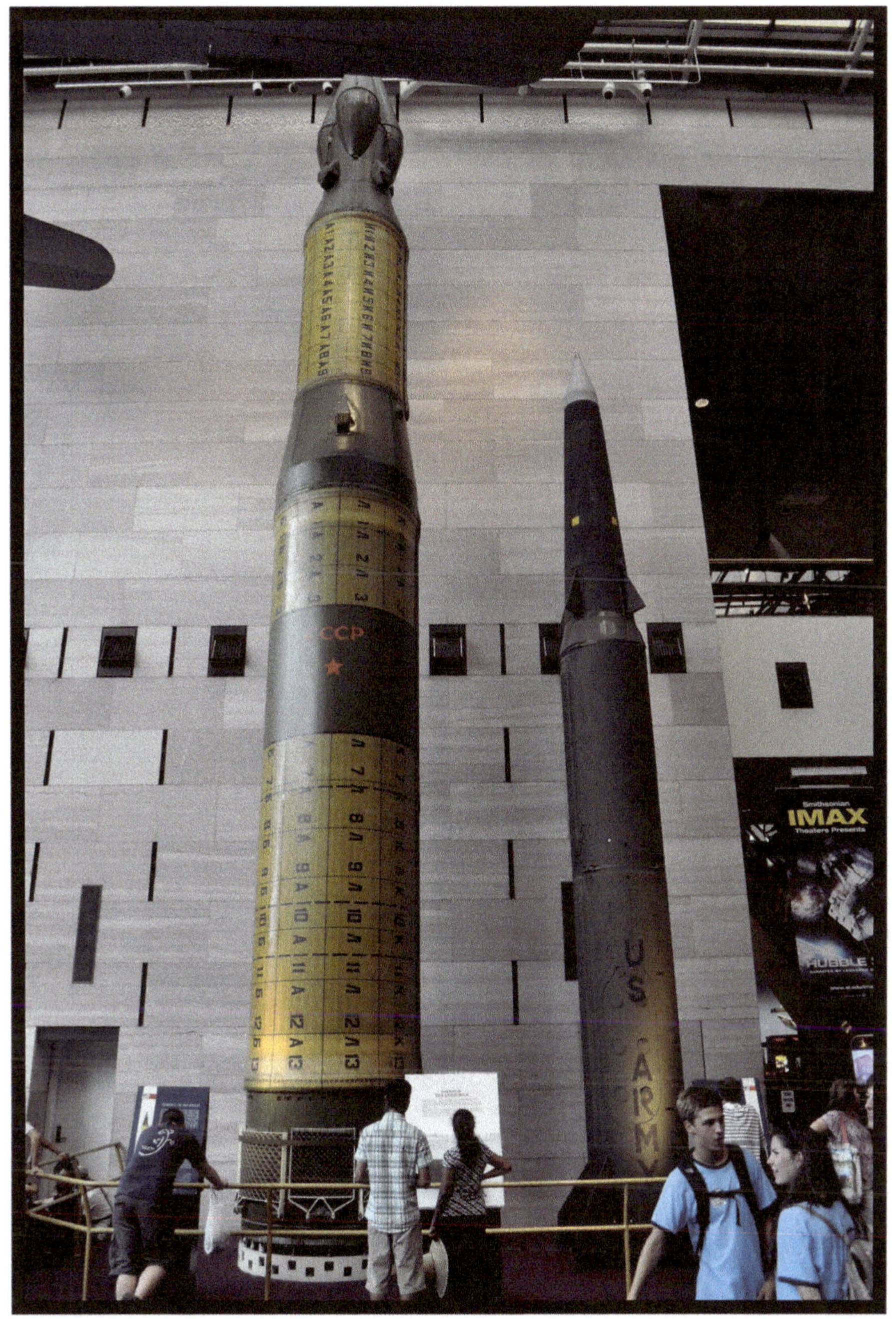

Side by Side. In the foreground, the Soviet SS-20 Intercontinental missile: behind it, the American Pershing-II ICBM. Both were designed to decapitate each other's civilizations. Now these Cold War relics are tourist attractions at the ***Smithsonian National Air and Space Museum, Washington, D.C., 2012***.

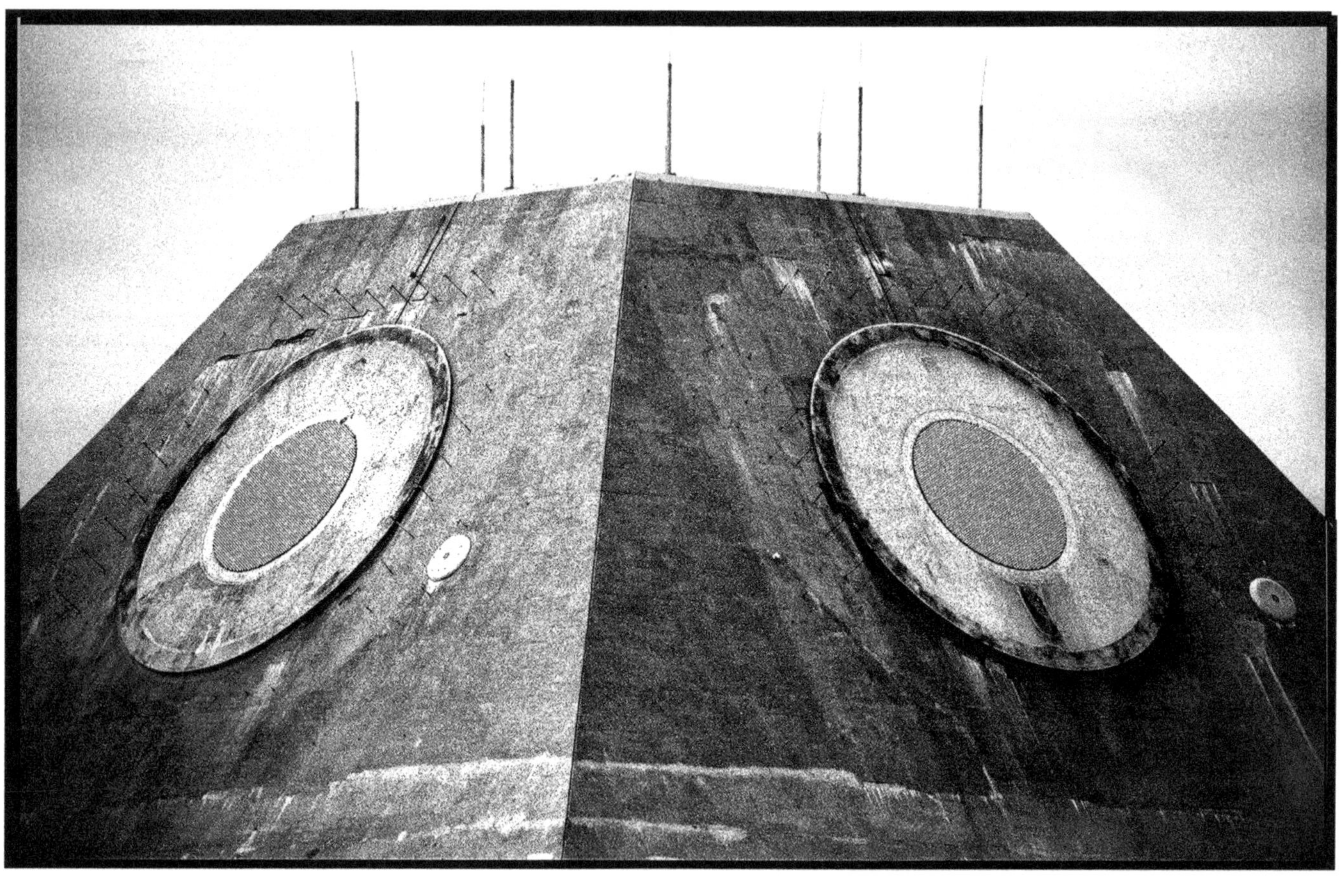

ABM Battle Management Radar. This self-contained radar facility was sited in the midst of an ICBM field. It was operational for only a short time before the Soviet Union and the U.S. signed a disarmament accord that resulted in its closure. *Nekoma, North Dakota, 1989.*

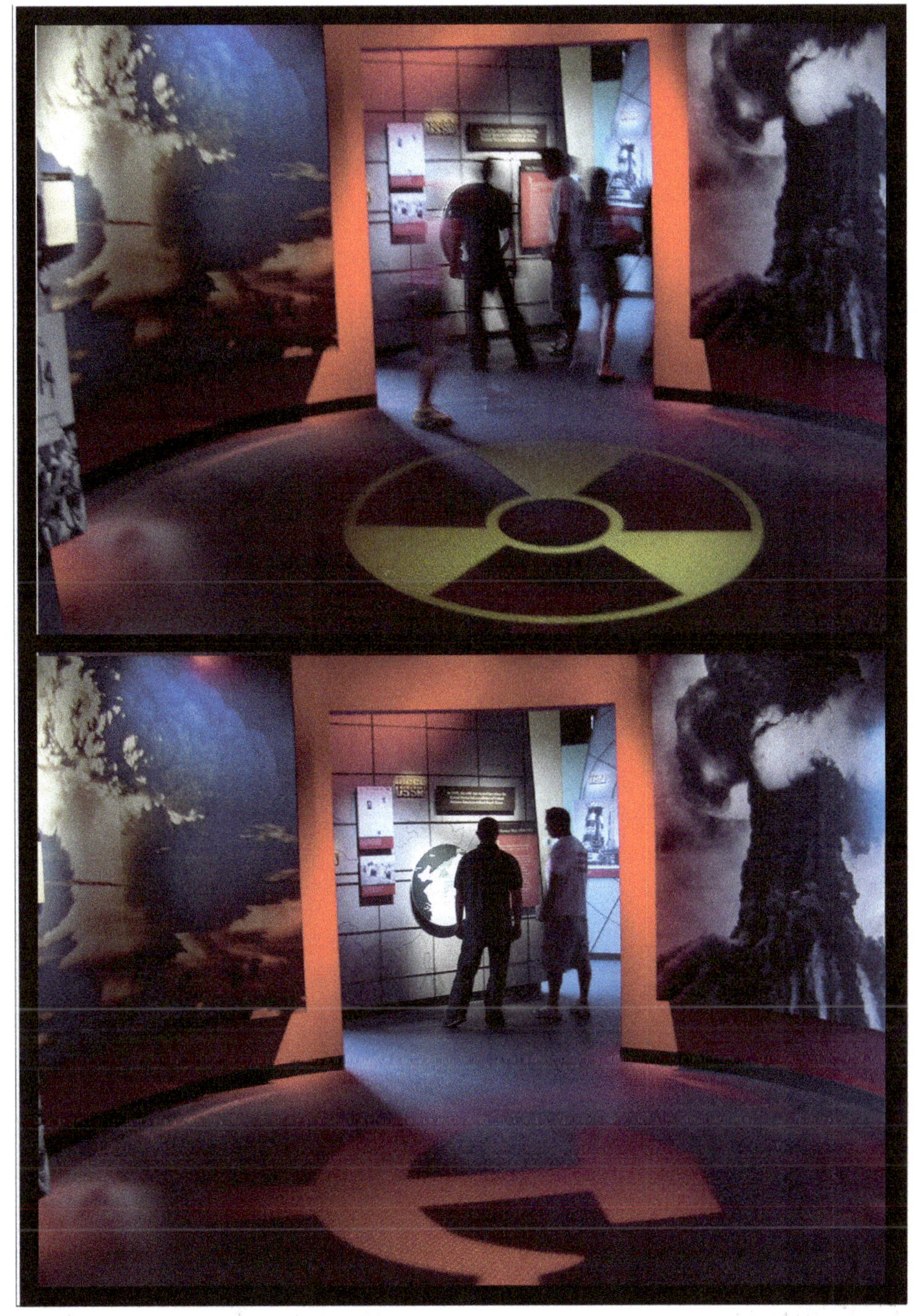

As Above, So Below. Above, an American Cold War display; below, a Soviet Cold War display. *Smithsonian Museum, Washington, D.C., 2004.*

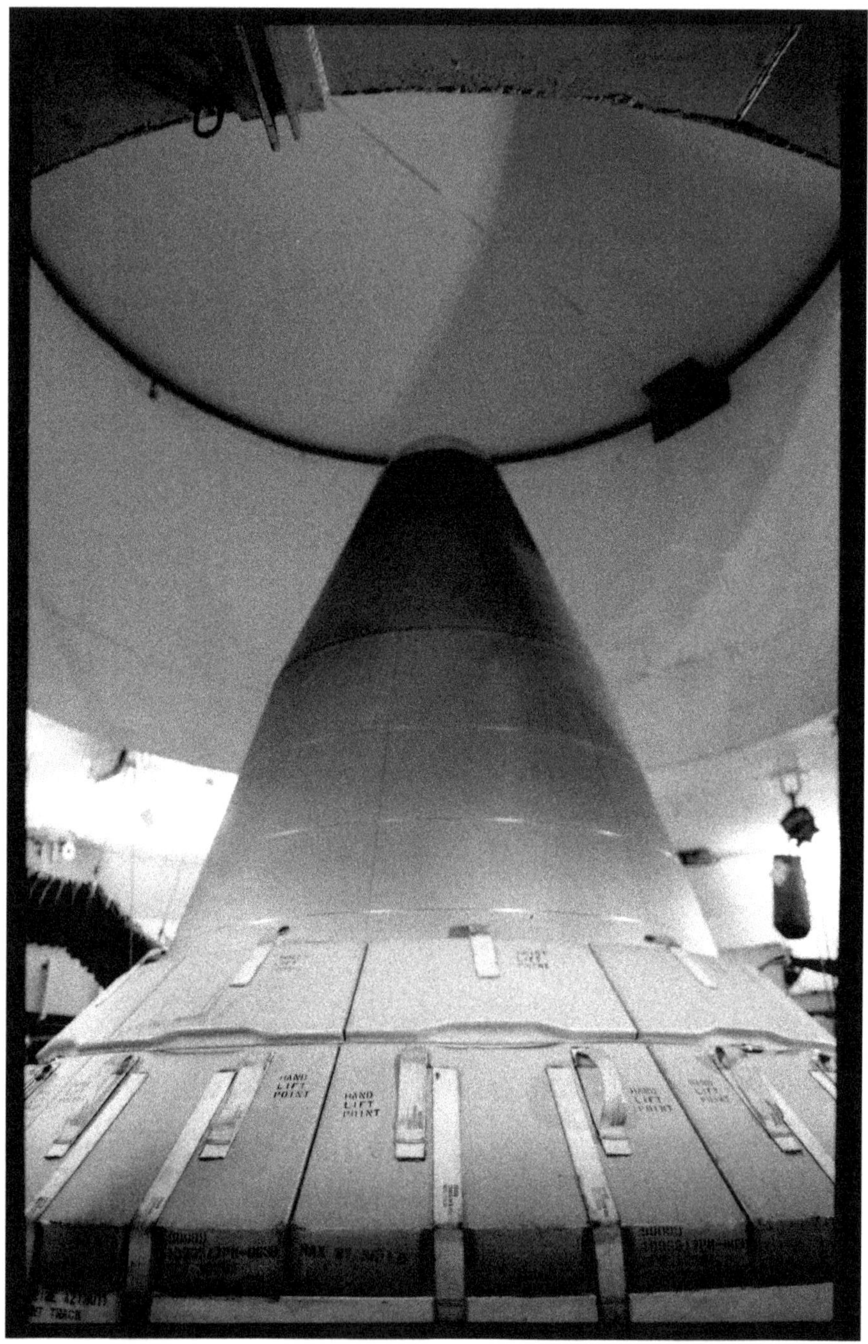

Peace-Keeper. For nearly 20 years, starting in 1986, the MX missile with its 10 independently targetable warheads was in the vanguard of the U.S. nuclear arsenal. *MX missile, Francis E. Warren Air Force Base, Cheyenne, Wyoming, 1989.*

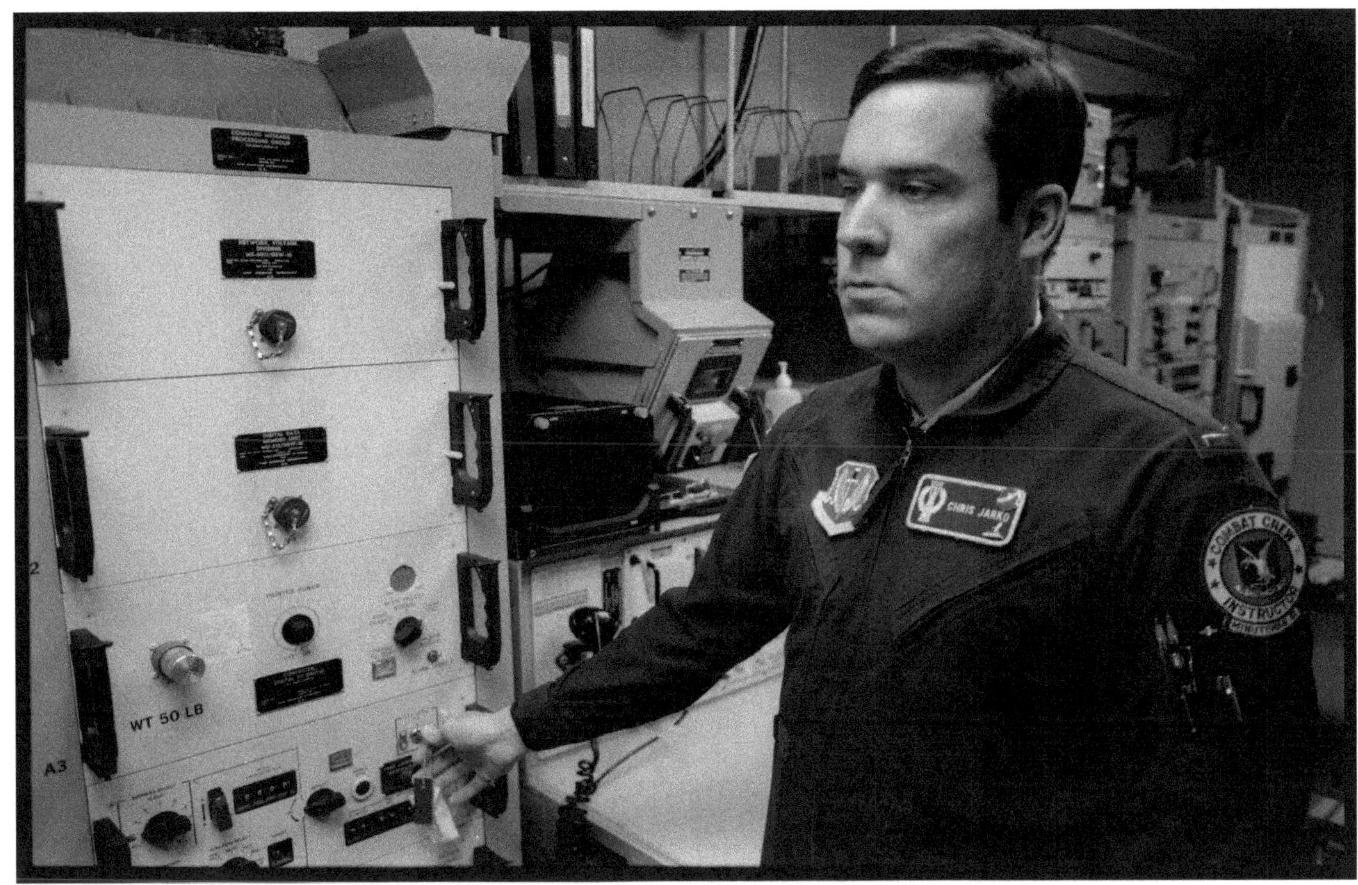

Missileer with Key. It takes two missileers, paces apart, turning two keys simultaneously, to launch a missile. This is a precaution to prevent a rogue officer from launching a missile on his own. *Francis E. Warren Air Force Base, Cheyenne, Wyoming, 1989.*

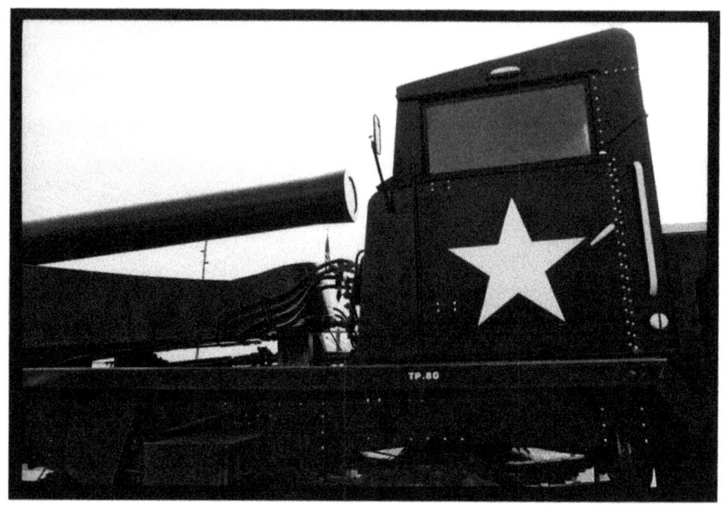

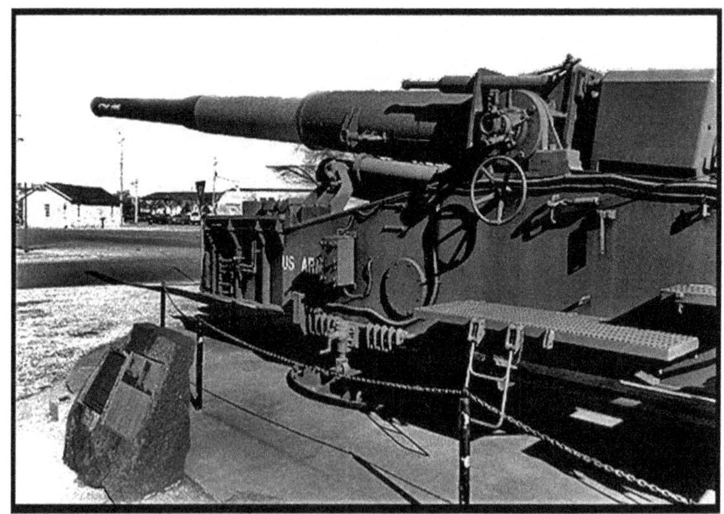

Atomic Cannon. *Kirtland Air Force Base, Albuquerque, New Mexico, 1988.*

"Atomic Annie" is the nickname for this cannon designed to deliver a tactical nuclear projectile within a twenty mile radius. It was primarily meant to defend the West from Soviet troops invading Germany through the Fulda Gap 60 miles northeast of Frankfurt. *Fort Sill, Oklahoma, 1995.*

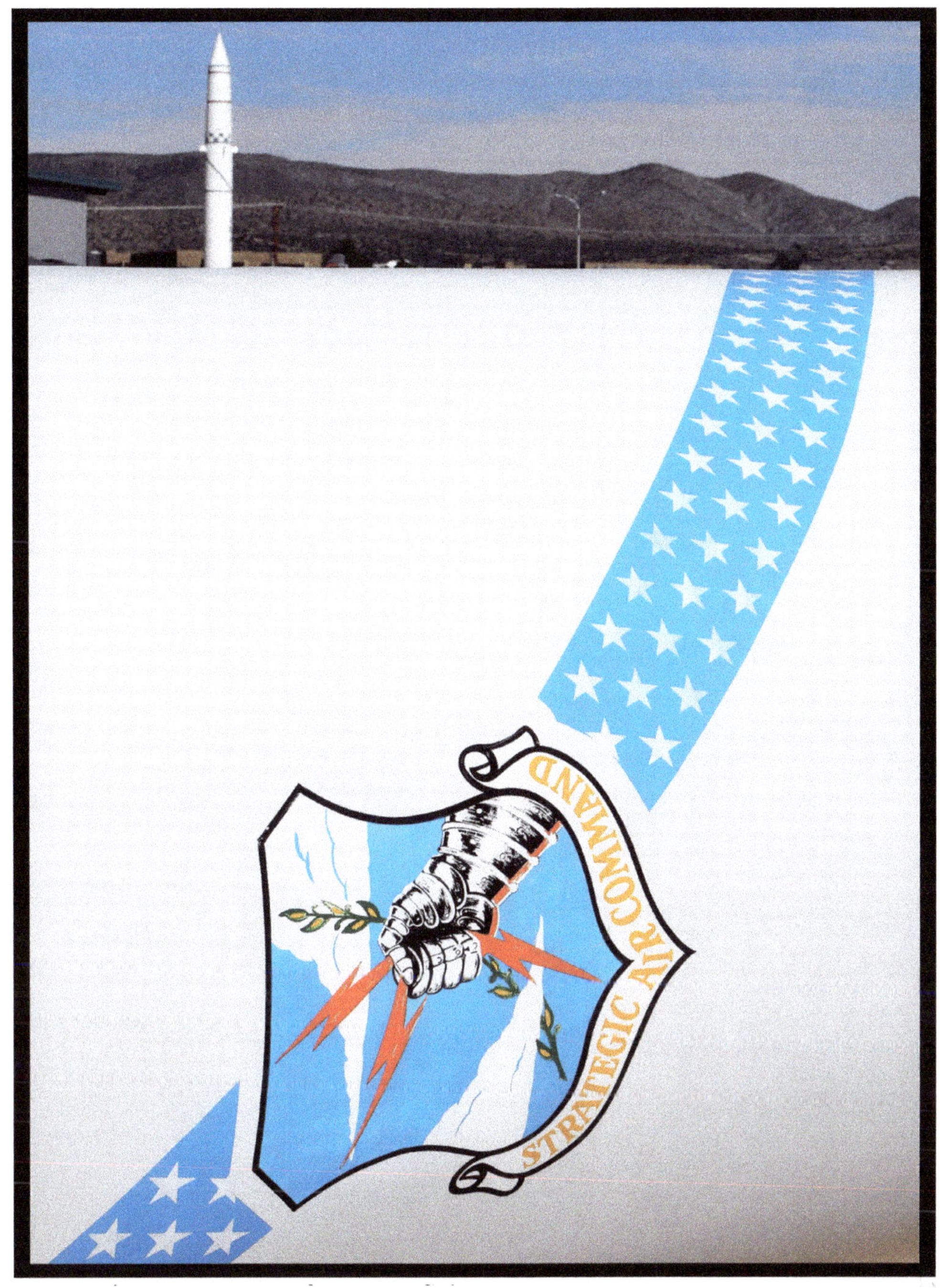

The National Museum of Nuclear Science and Industry exhibits deactivated nuclear missiles. It is affiliated with the Smithsonian Institute in Washington, D.C. *Albuquerque, New Mexico, 2013.*

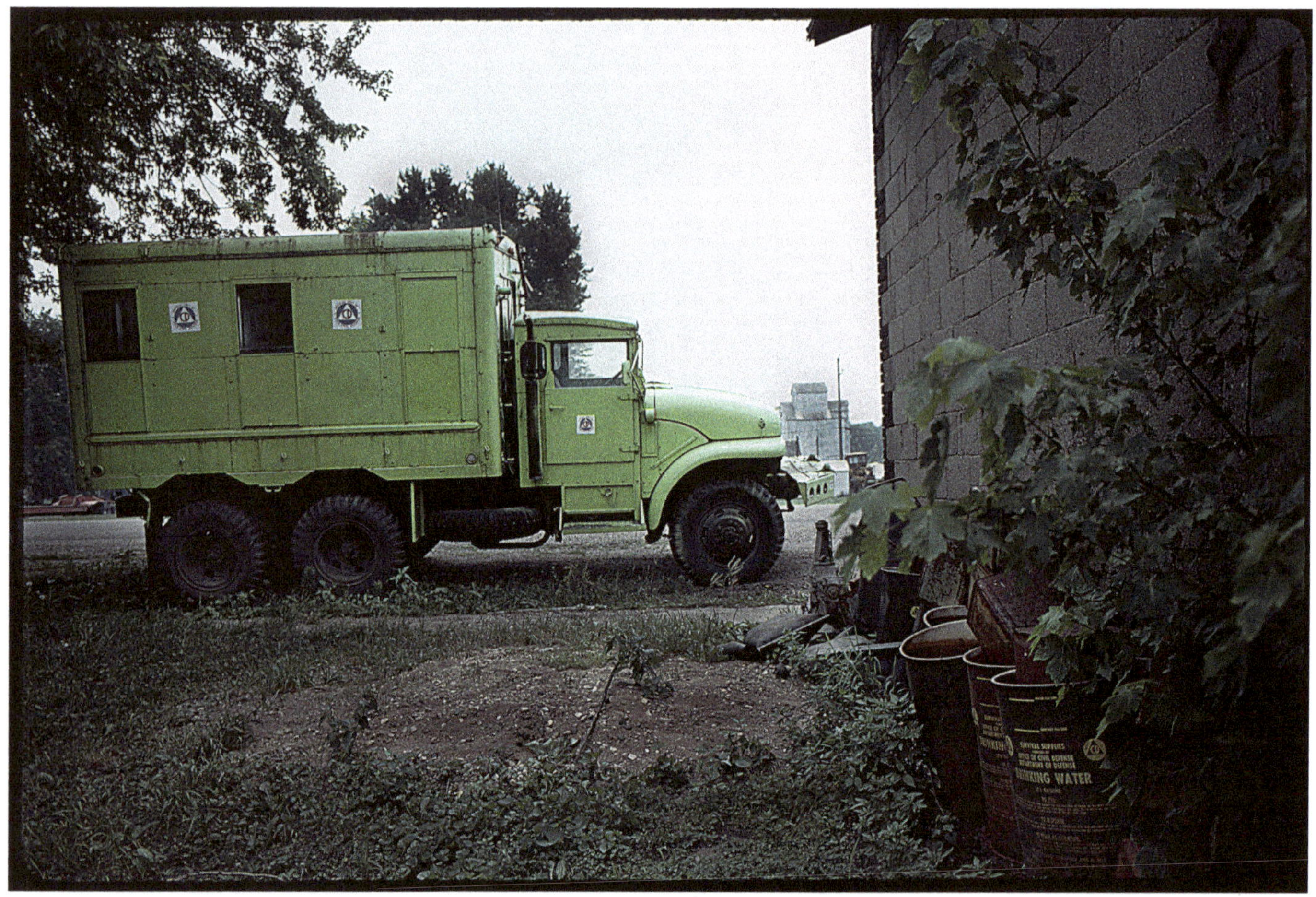

Iowa Civil Defense Post-Attack Resource Vehicle meant to distribute first aid, medicines, and other resources to survivors attempting to hold their own in farmland shelters and home basements after a nuclear attack. How vehicle drivers would protect themselves as they navigated radioactive terrain between homesteads is not known, nor has my research unearthed data on post-apocalyptic food, medicine, and fuel depots for the region. Long-abandoned Civil Defense supply canisters litter the ground. *Northwestern Iowa, 1985.*

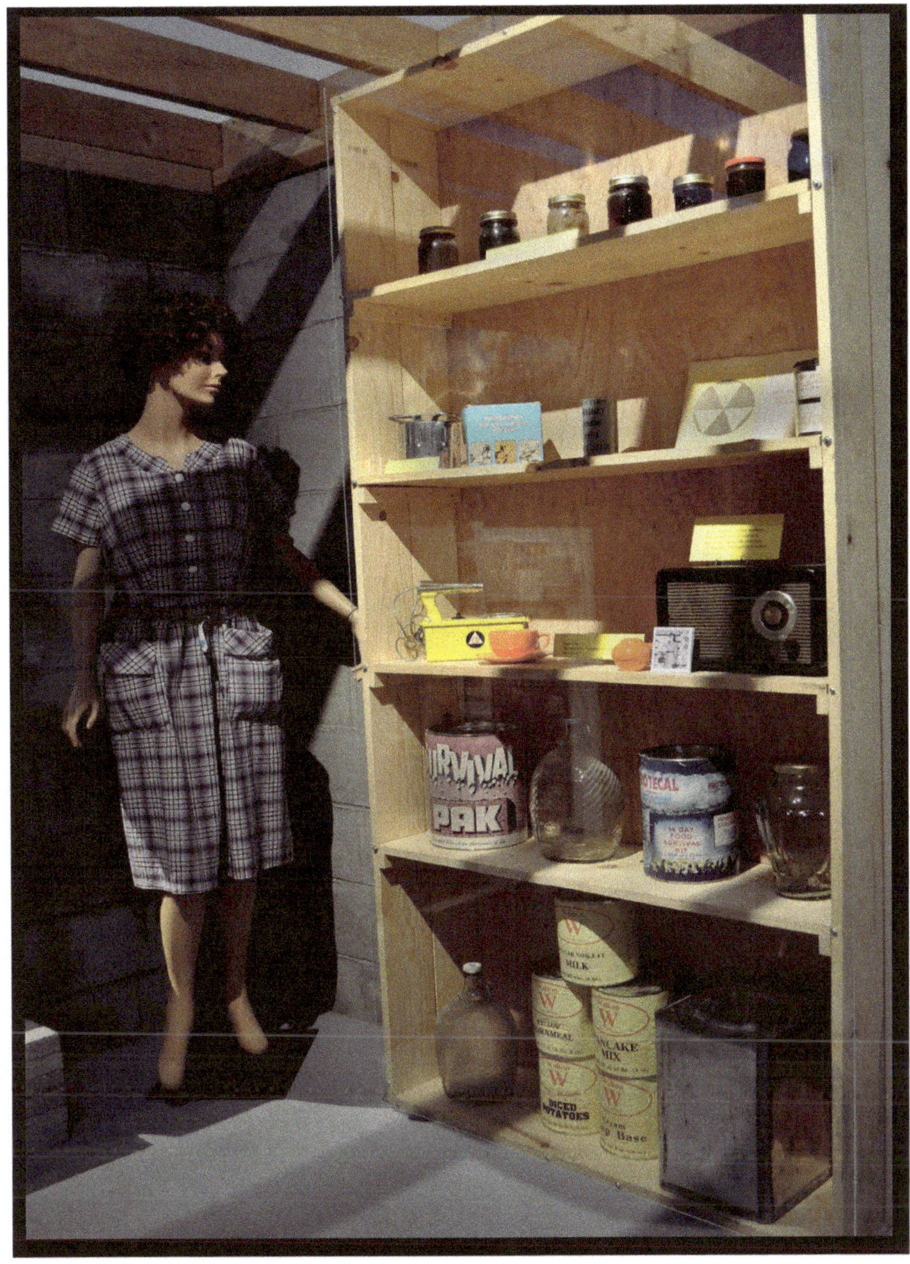

Model Housewife with Modern Fallout Shelter Survival Cupboard.
National Atomic Testing Museum, Las Vegas, Nevada, 2011.

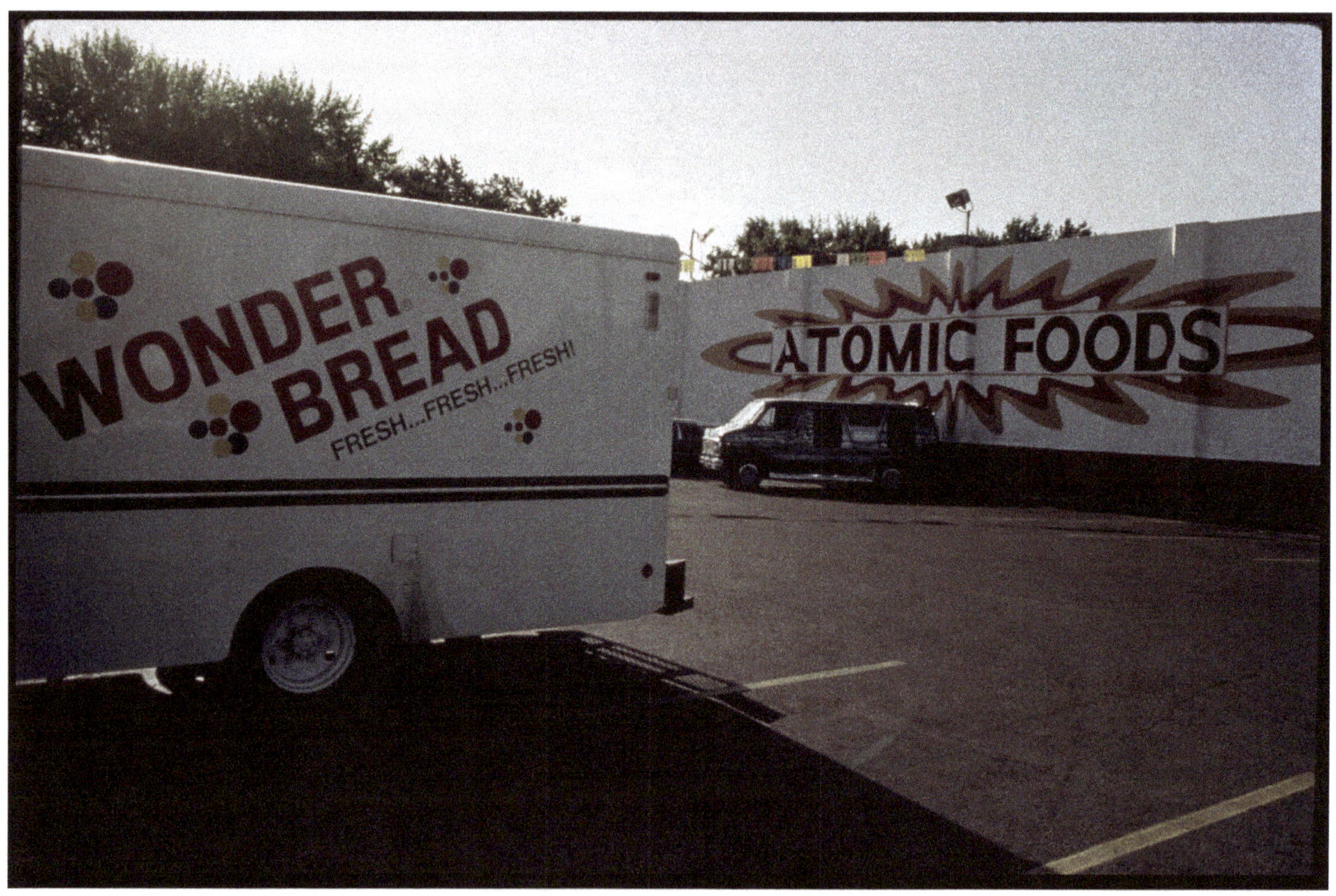

Wonder Bread at Atomic Foods near the Hanford Nuclear Reservation. Hanford created plutonium for most of the warheads in the US nuclear arsenal. *Pasco, Washington, 1993.*

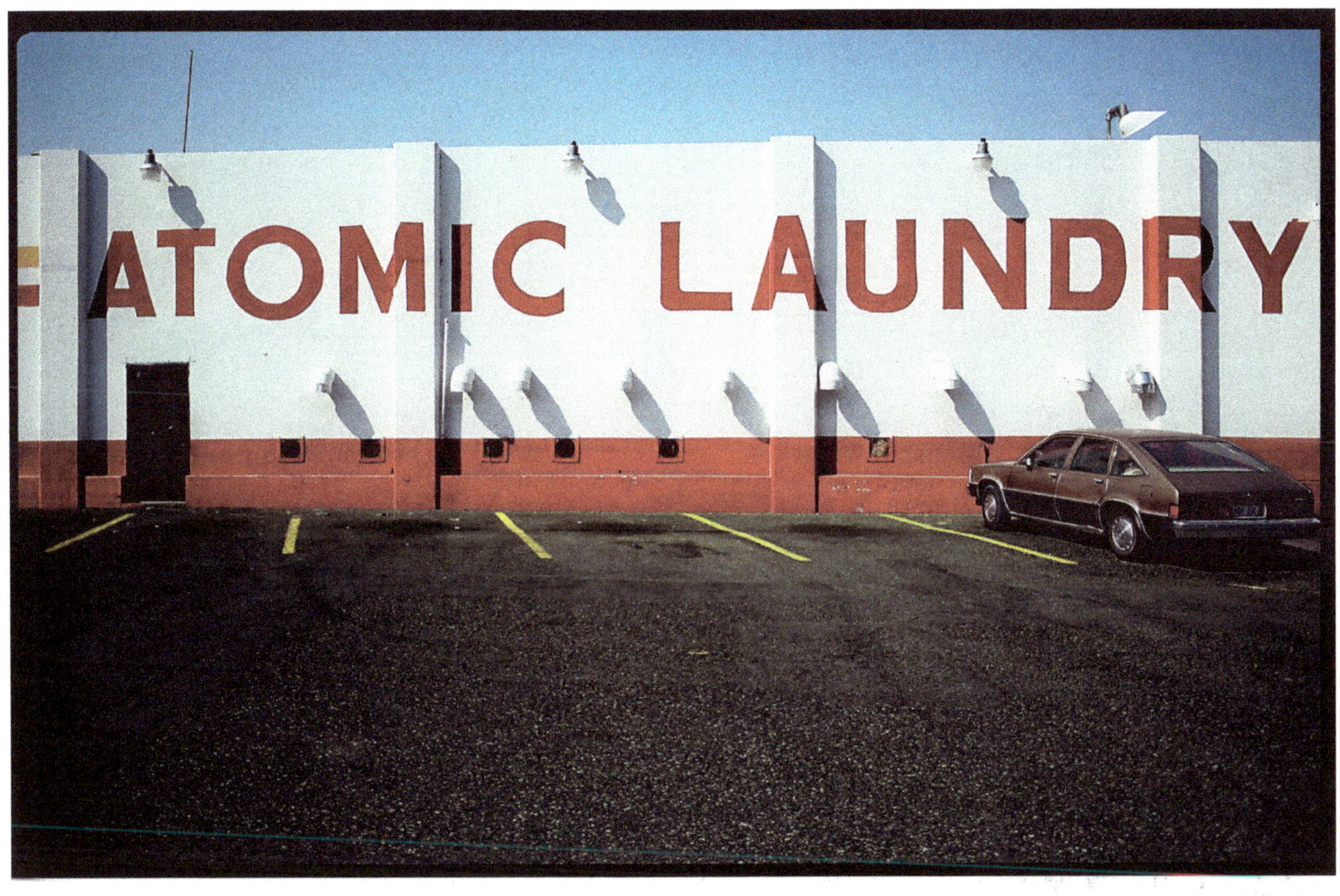

The Atomic Laundry is patronized by workers and their families living near the Hanford Nuclear Reservation. *Pasco, Washington, 1993.*

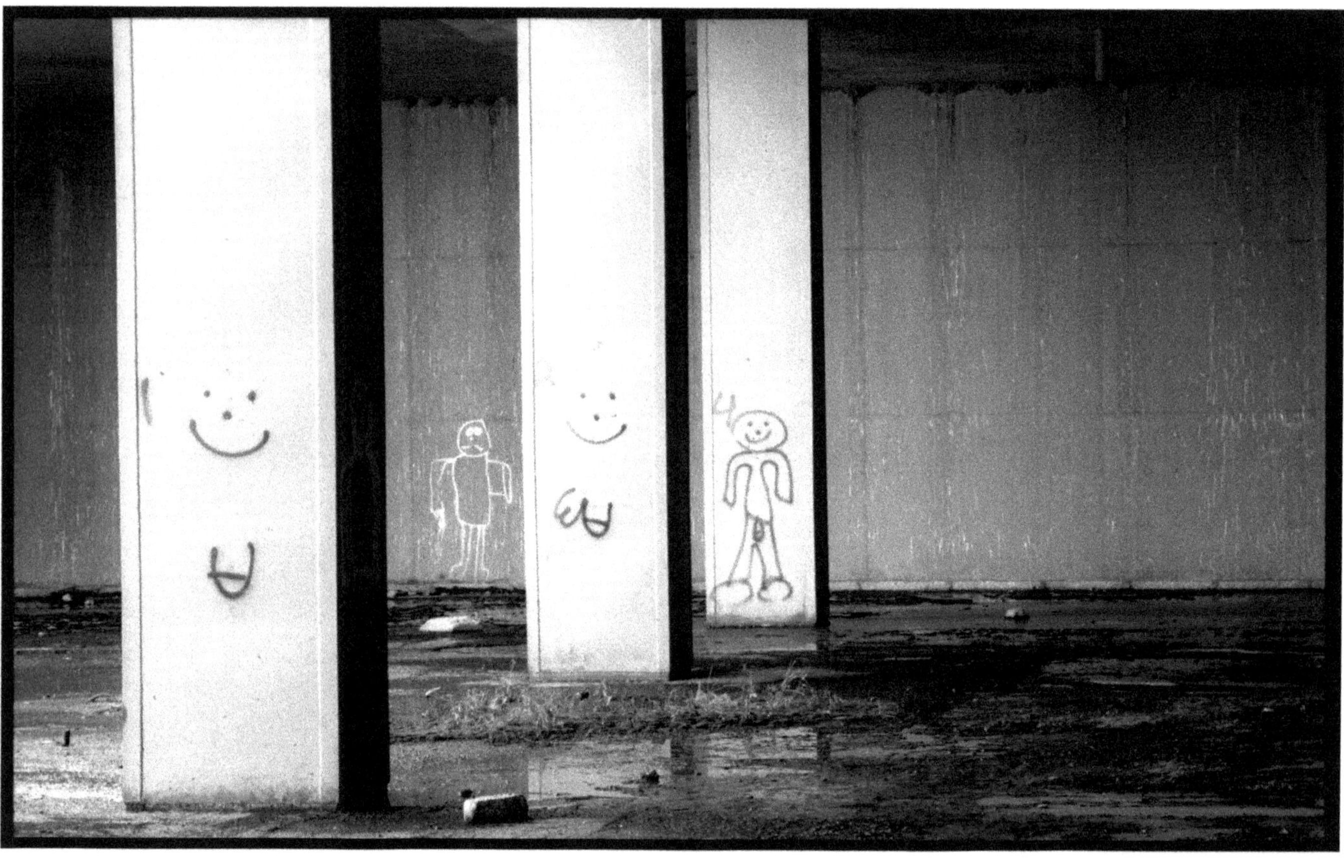

Anti-Ballistic Radar Support Columns for a facility designed to detect incoming enemy missiles and activate Spartan long range missiles to destroy them in the upper atmosphere, or quicker short-range Sprints to destroy missiles that got past the Spartans. Construction on the site was terminated in 1974 after the signing of the 1972 Salt 1 treaty between Nixon and Brezhnev. *Shelby, Montana, 1993.*

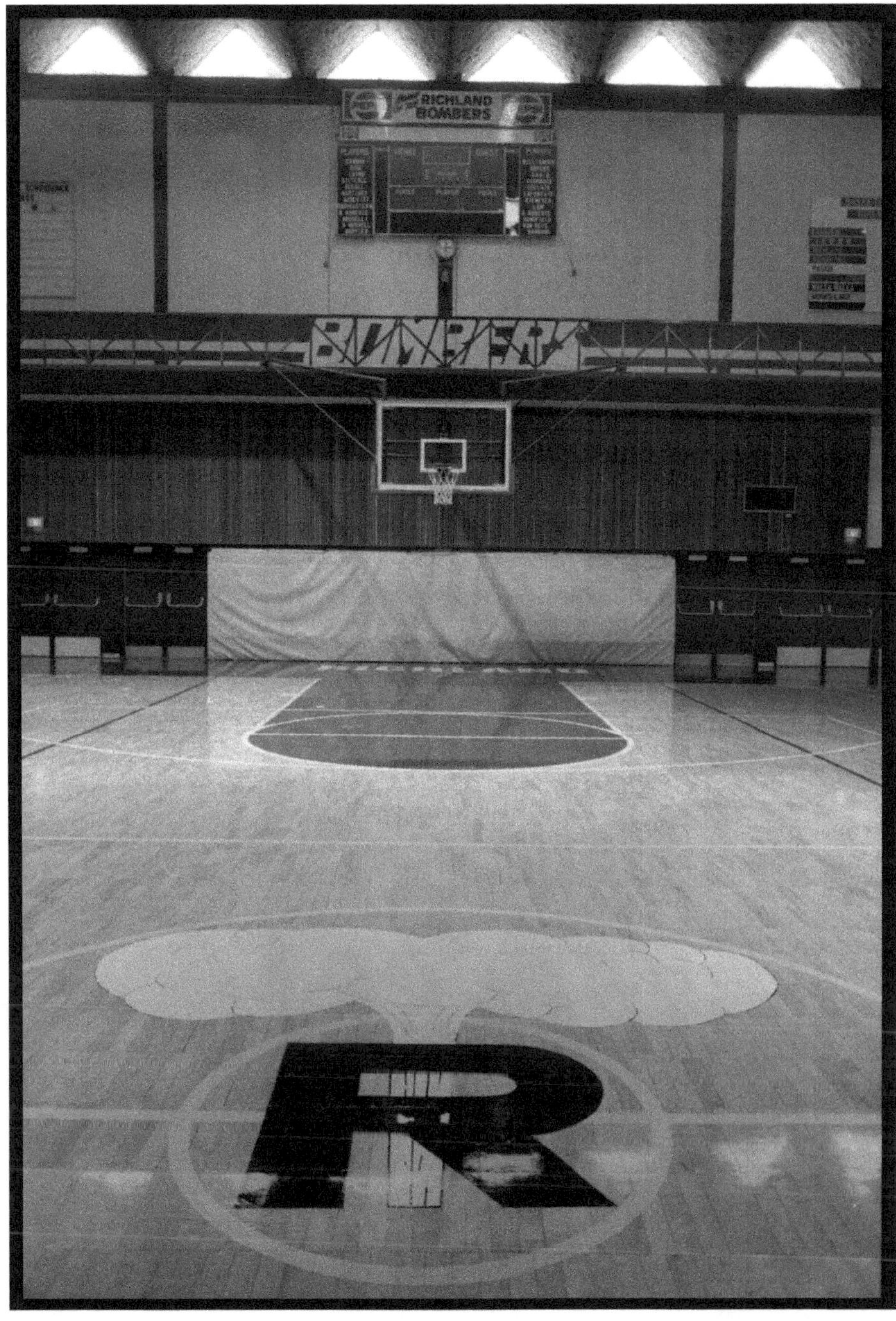

Basketball for Bombers. In spite of controversy, the mushroom cloud has long remained the beloved logo of the Richland Bombers—a high school basketball team ten miles from the Hanford Plutonium Production Reactors. *Richland, Washington, 1993.*

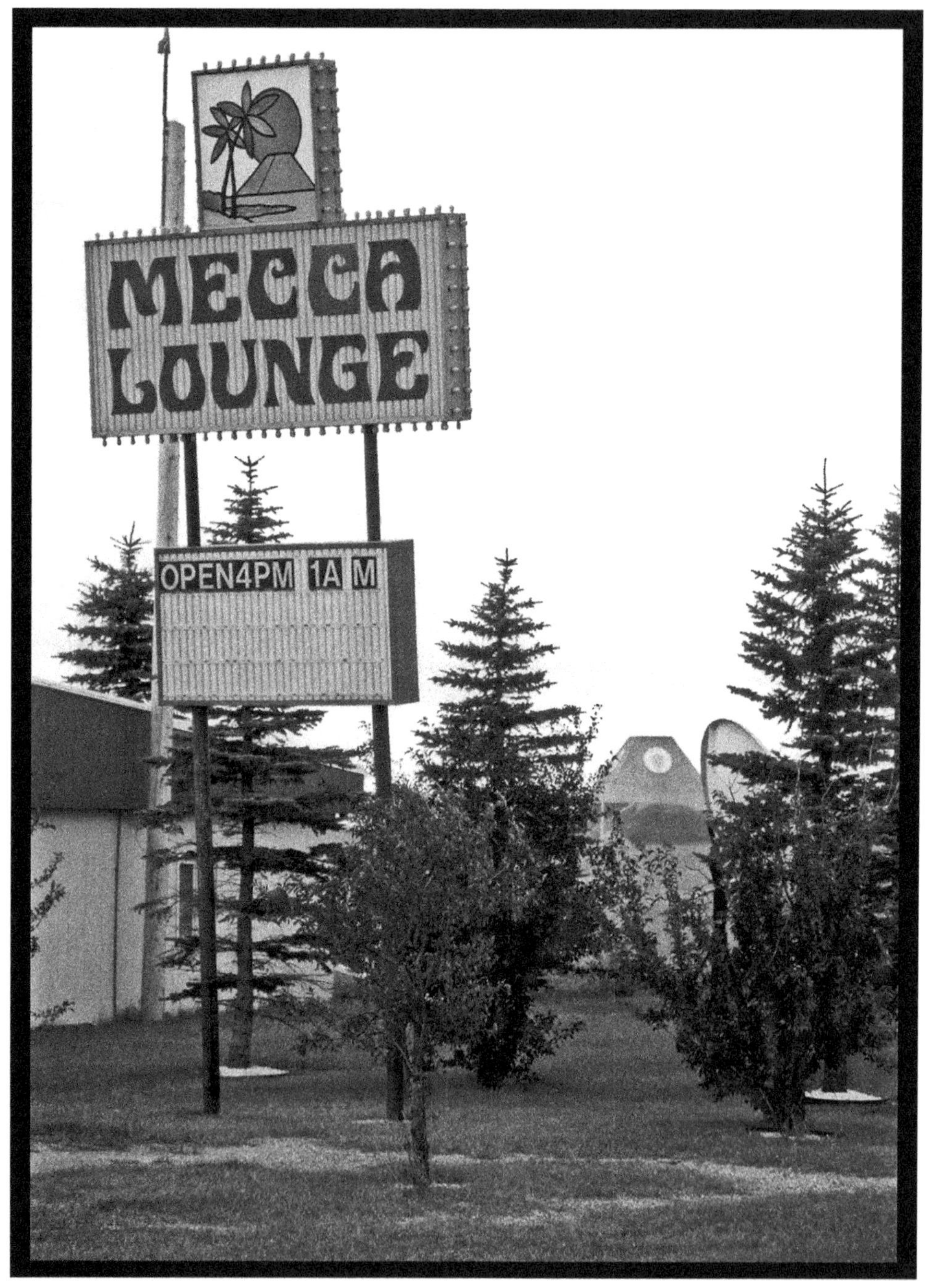

Mecca Lounge. A short distance from a de-activated ABM radar facility, this cocktail lounge bases its name on a religious shrine in a dry country. *Nekoma, North Dakota, 1989.*

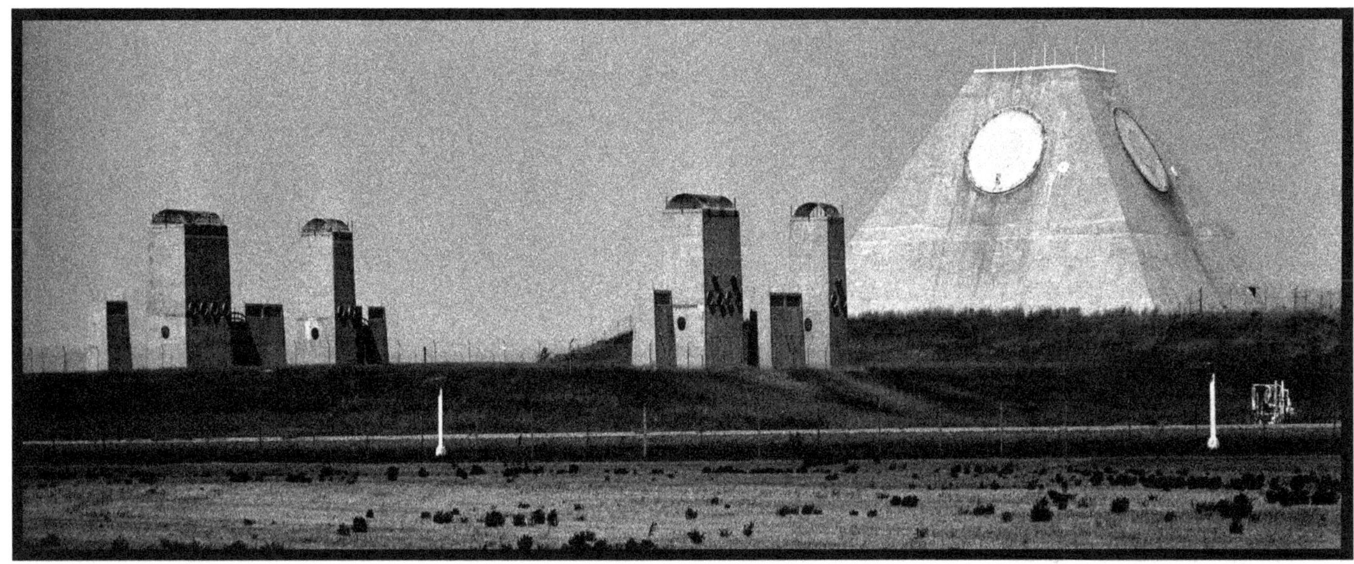

ABM Battle Management Radar and Missile Field were active for only one day in 1975 before a nuclear agreement between the Soviet Union and the U.S. forced its closure. *Nekoma, North Dakota, 1989.*

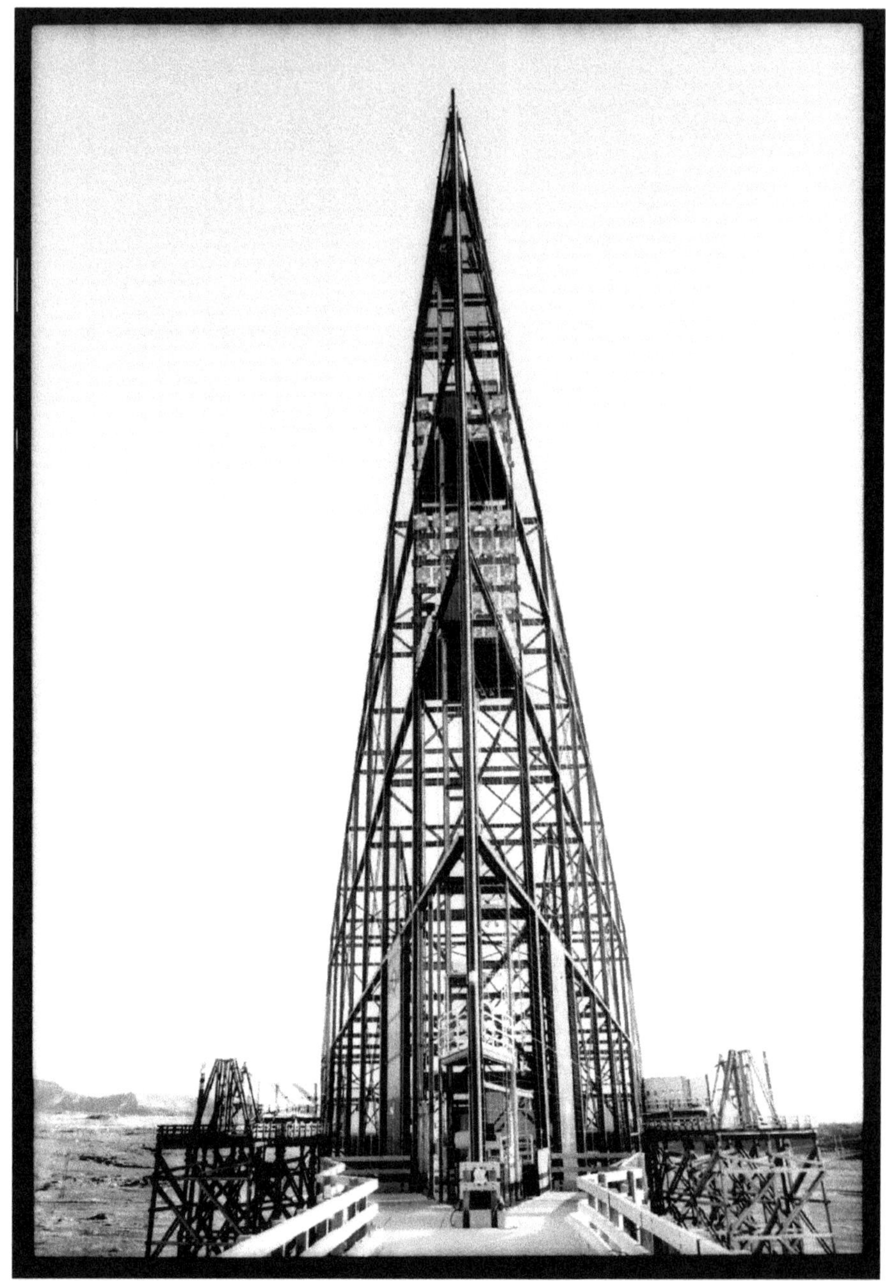

Electromagnetic Pulse Tower. A nuclear explosion in space emits a pulse that can wipe out large portions of the U.S. power grid. This facility has been built without any metal parts to simulate an airborne environment for aircraft and other electronic devices in order to battle-harden them to the potent effects of the pulse. *Kirtland Air Force Base, Albuquerque, New Mexico, 1994.*

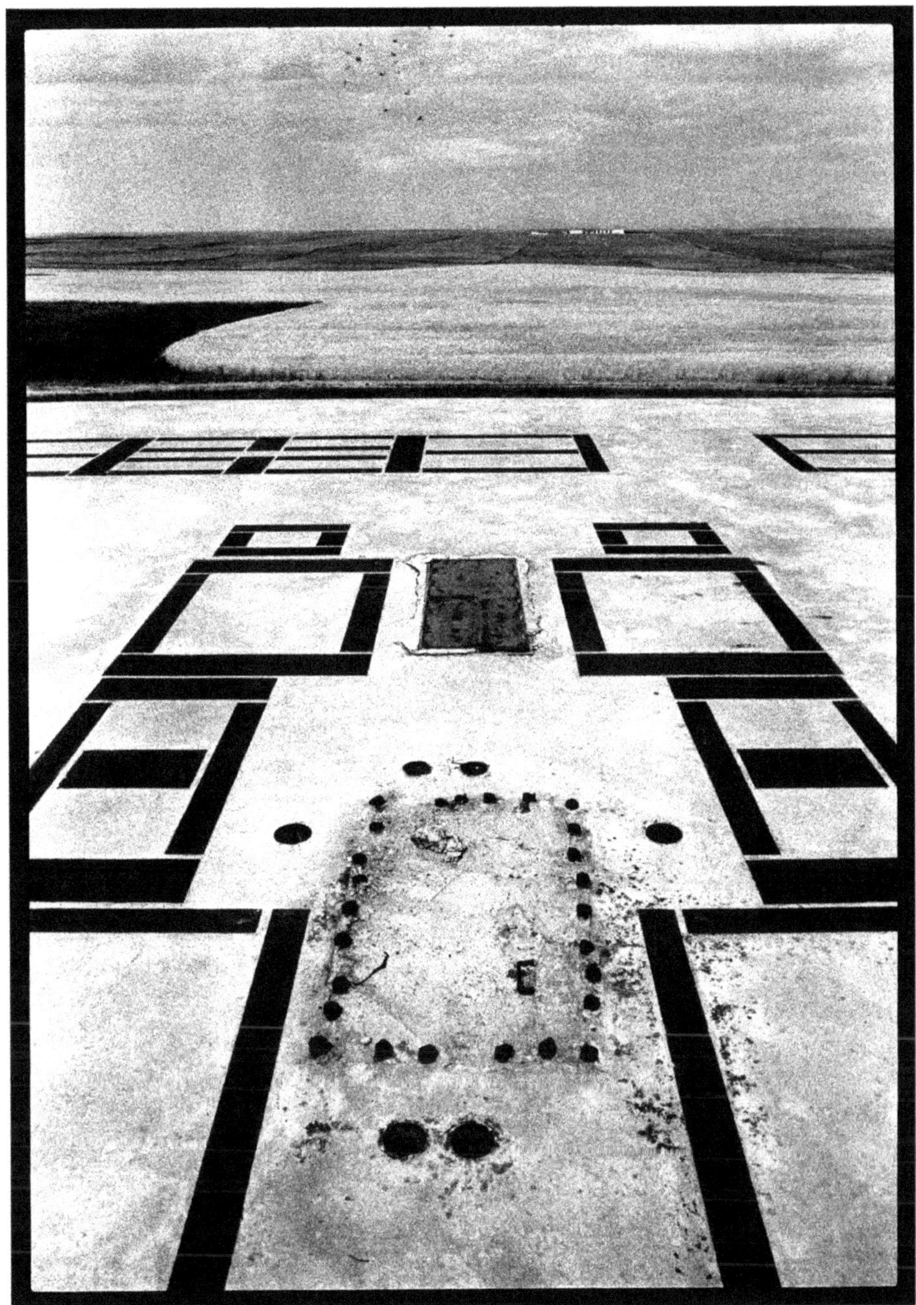

ABM Battle Management Radar Facility Under Construction. Twenty feet above wheat fields where protective missile fields were planned, grid-like outlines mark off an area designed to exist inside a pyramid-shaped radar facility. The SALT 1 Agreement between the Soviet Union and the U.S. halted construction in 1972. *Shelby, Montana, 1989.*

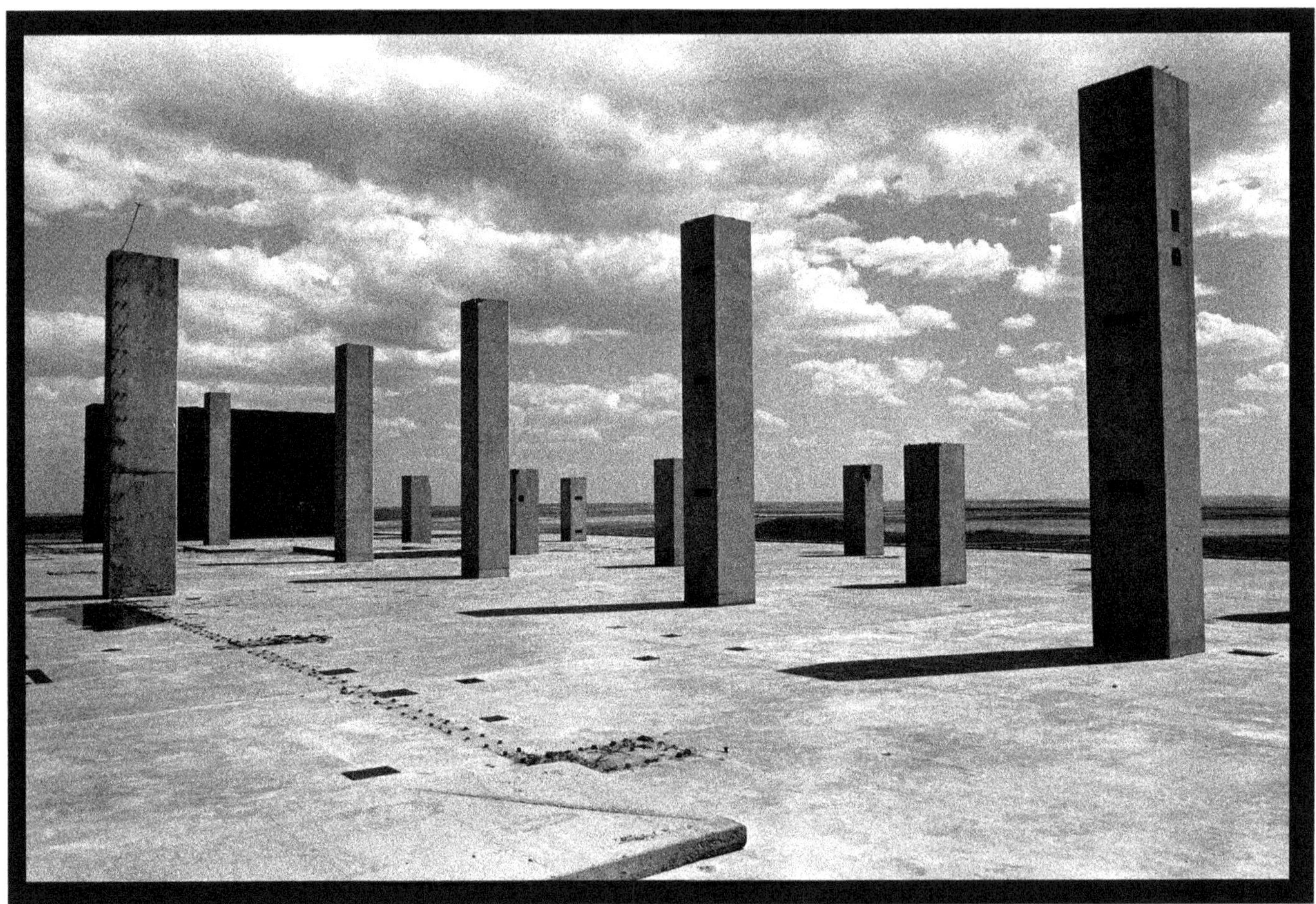

Field of Monoliths is part of an unfinished pyramid-shaped radar facility to monitor missiles attacking the U.S. The 30 long-range Spartan missiles and 70 short-range, faster Sprints assigned to the facility were deemed inadequate in the face of incoming, multiple-independently-targetable Soviet missiles capable of splitting their ten warheads into ten different trajectories aimed at ten indefensible targets. *Shelby, Montana, 1989.*

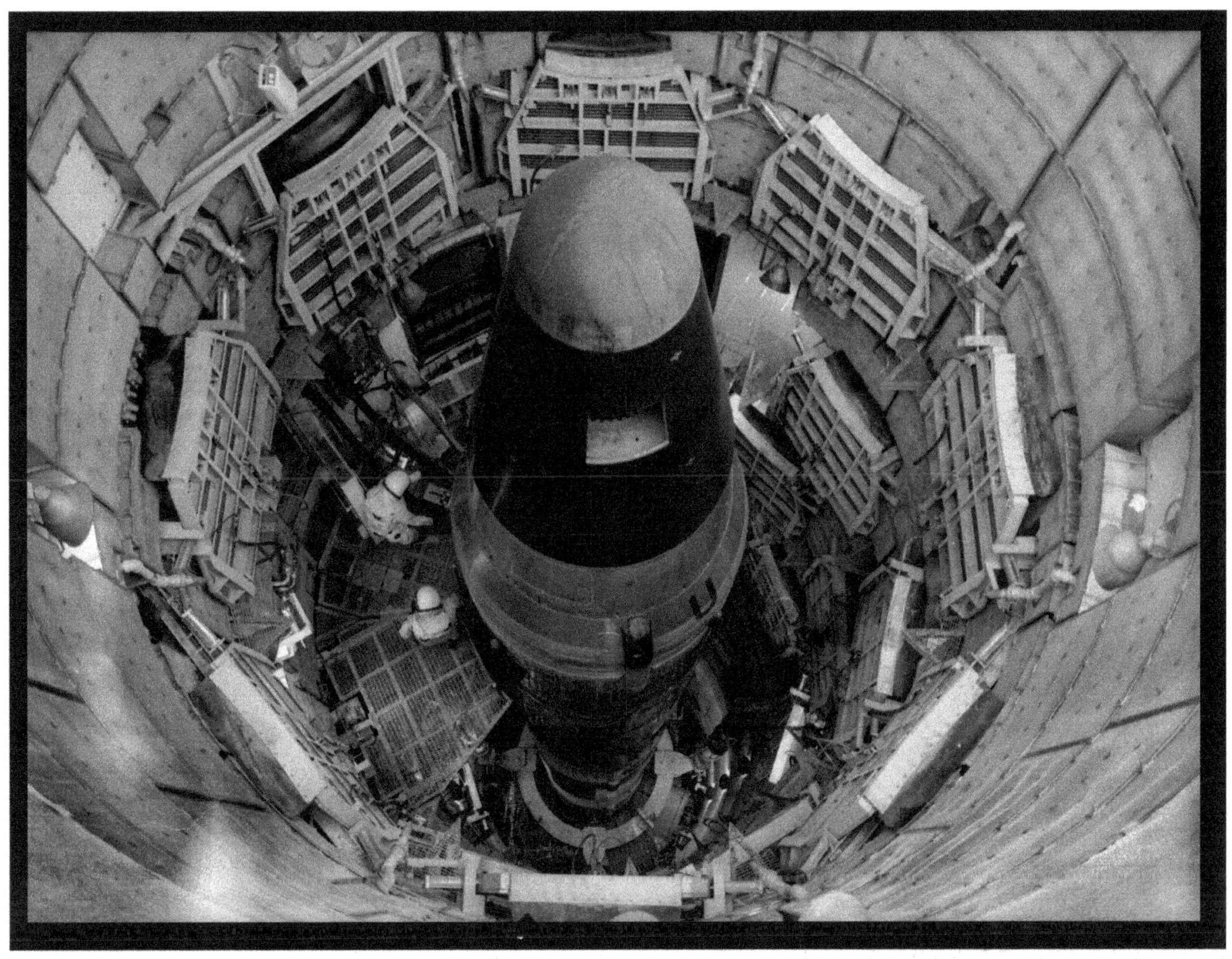

Removing a Section of the Nosecone of this Titan missile allowed Soviet satellites to confirm that the missile had indeed been de-activated. *Green Valley Titan Missile Museum, Green Valley, Arizona, 2003.*

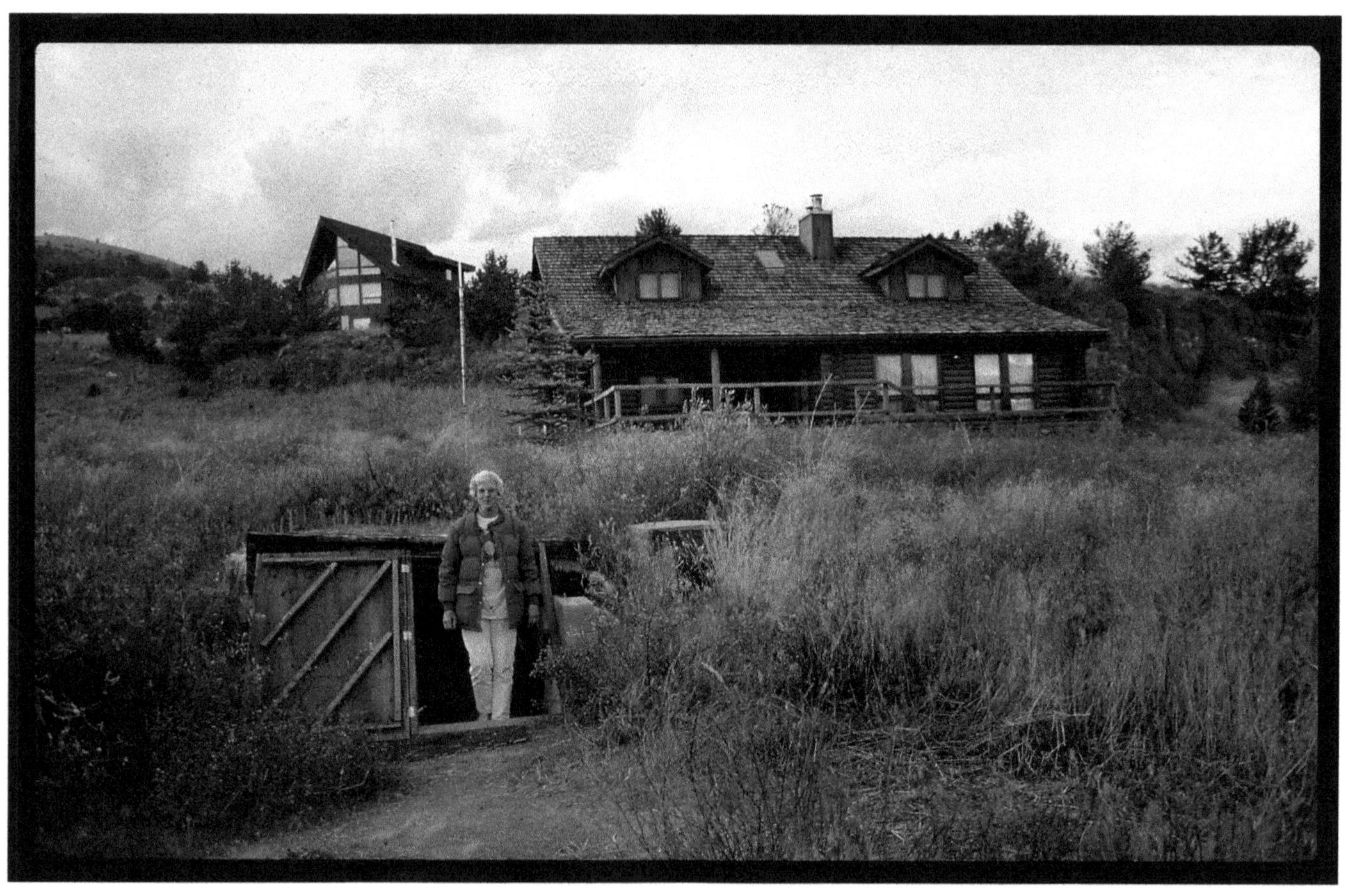

Eleanor Shiefflin at the Entry to the Community Bomb Shelter for the Church Universal and Triumphant. Church founder Elizabeth Clare Prophet believed in surviving nuclear Armageddon. Her devotees constructed this shelter with a communal kitchen and private bedrooms with bunk beds. *Glastonbury, Montana, 1993.*

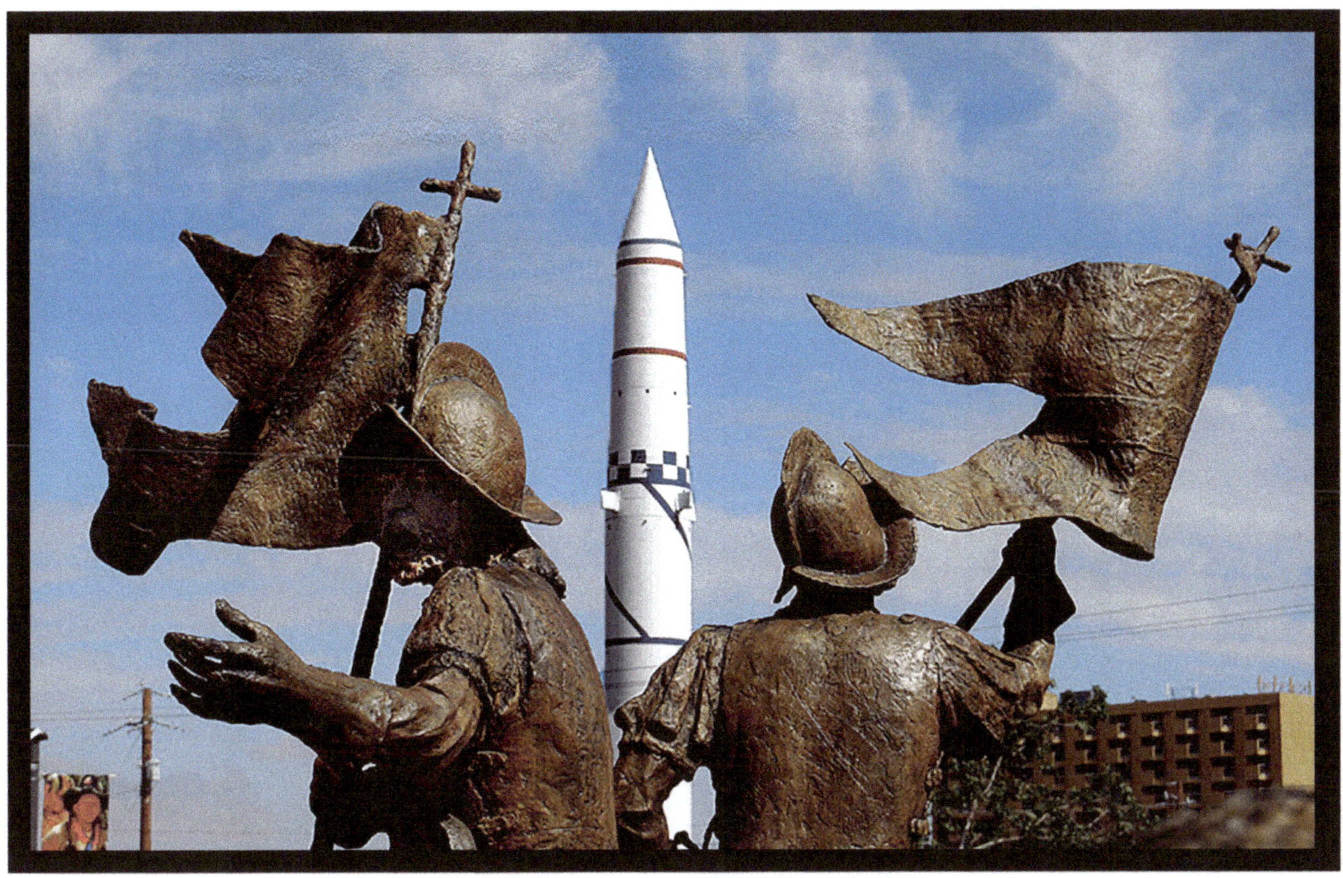

Conquistadors and the Bomb. Hungry for gold, land, and converts, Spanish Conquistadors overran South America, Florida, and the American Southwest from the 15th to 17th centuries. At Albuquerque's Art and History Museum across from the National Atomic Museum, two warring cultures meet. Sighting a machined perfection he has never seen before, a conquistador signals a warning to fellow conquerors. New Mexico has from Day One of the nuclear age been colonized by the culture of nuclear weapons. *New Mexico Museum of Art and History opposite the National Atomic Museum, 2006.*

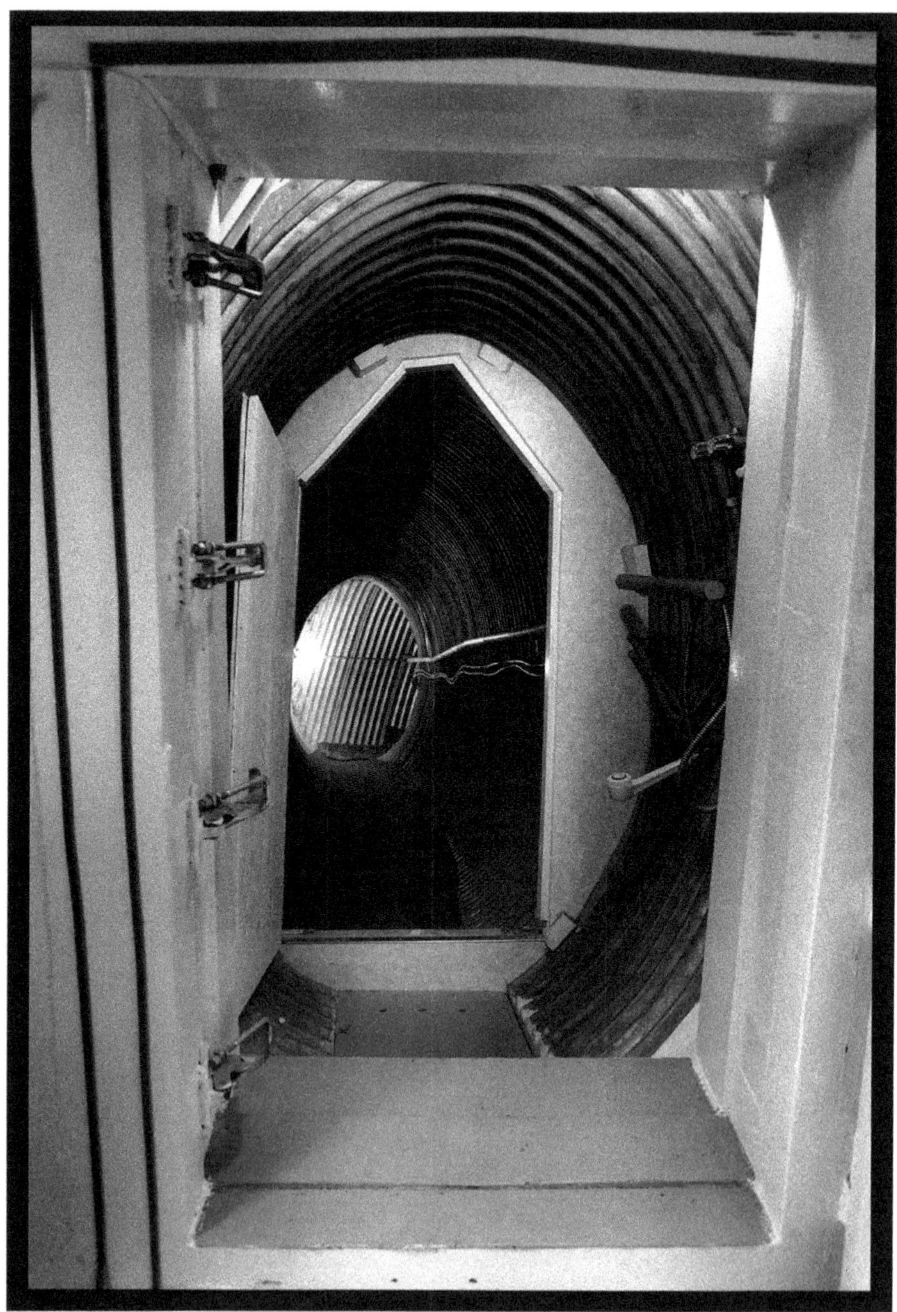

Egg/Coffin. Elizabeth Clare Prophet's nuclear survivors were required to shower before passing through this entrance to their church community shelter. Once cleansed of radioactive particles, they would settle deeper underground to outwait the deleterious effects of atomic fallout on their land. *Glastonbury, Montana, 1993.*

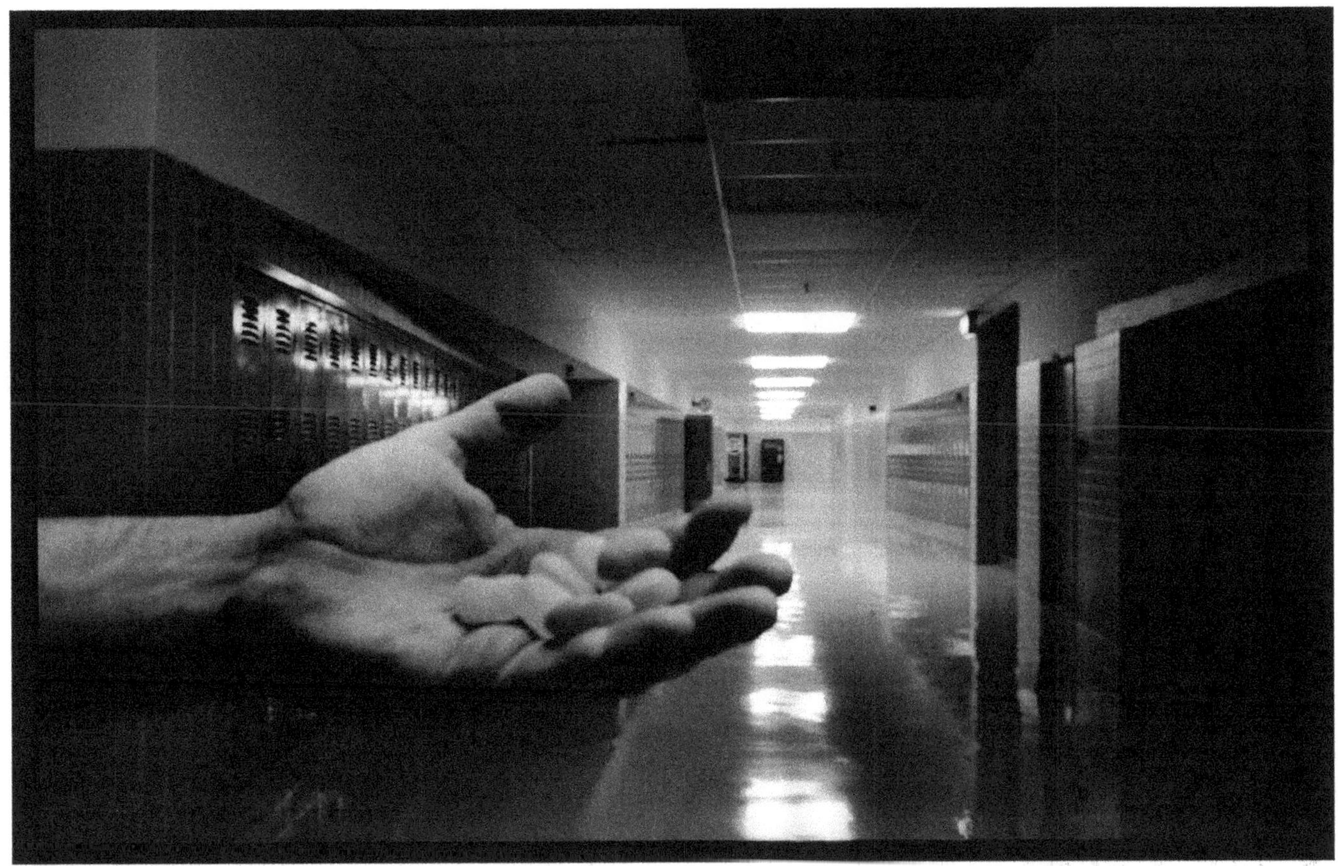

Long-lived Shelter Candies await victims, survivors, and tourists entering the underground Artesia High School bomb shelter. *Artesia, New Mexico, 1994.*

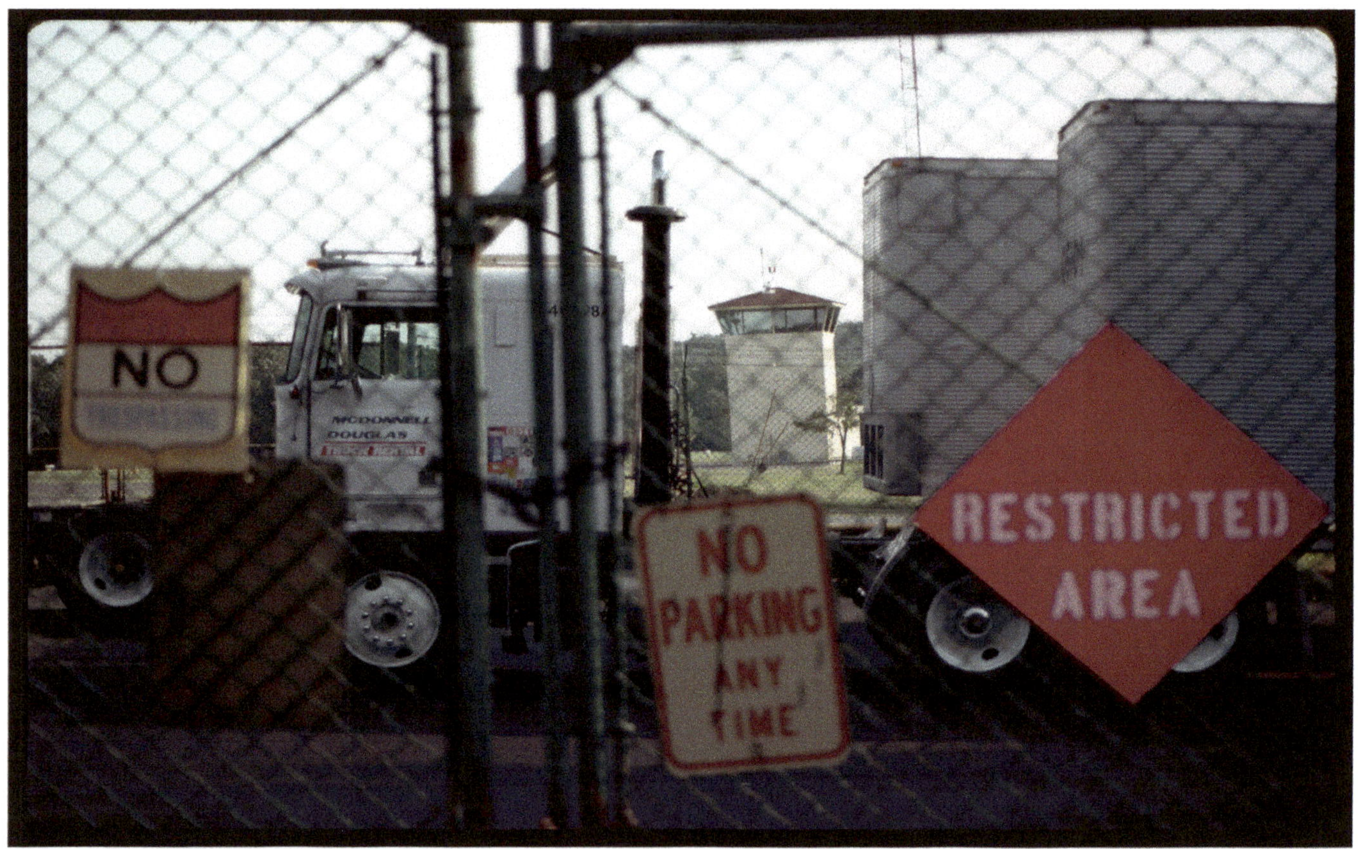

Mount Weather Presidential Shelter is meant to provide continuity for the President of the United States and his key officials. Space within is limited; so wives and children, as well as pilots and drivers, would likely be excluded. A direct hit by one of today's more powerful nuclear warheads would make survival inside this facility moot. *Mount Weather, Virginia, 2003.*

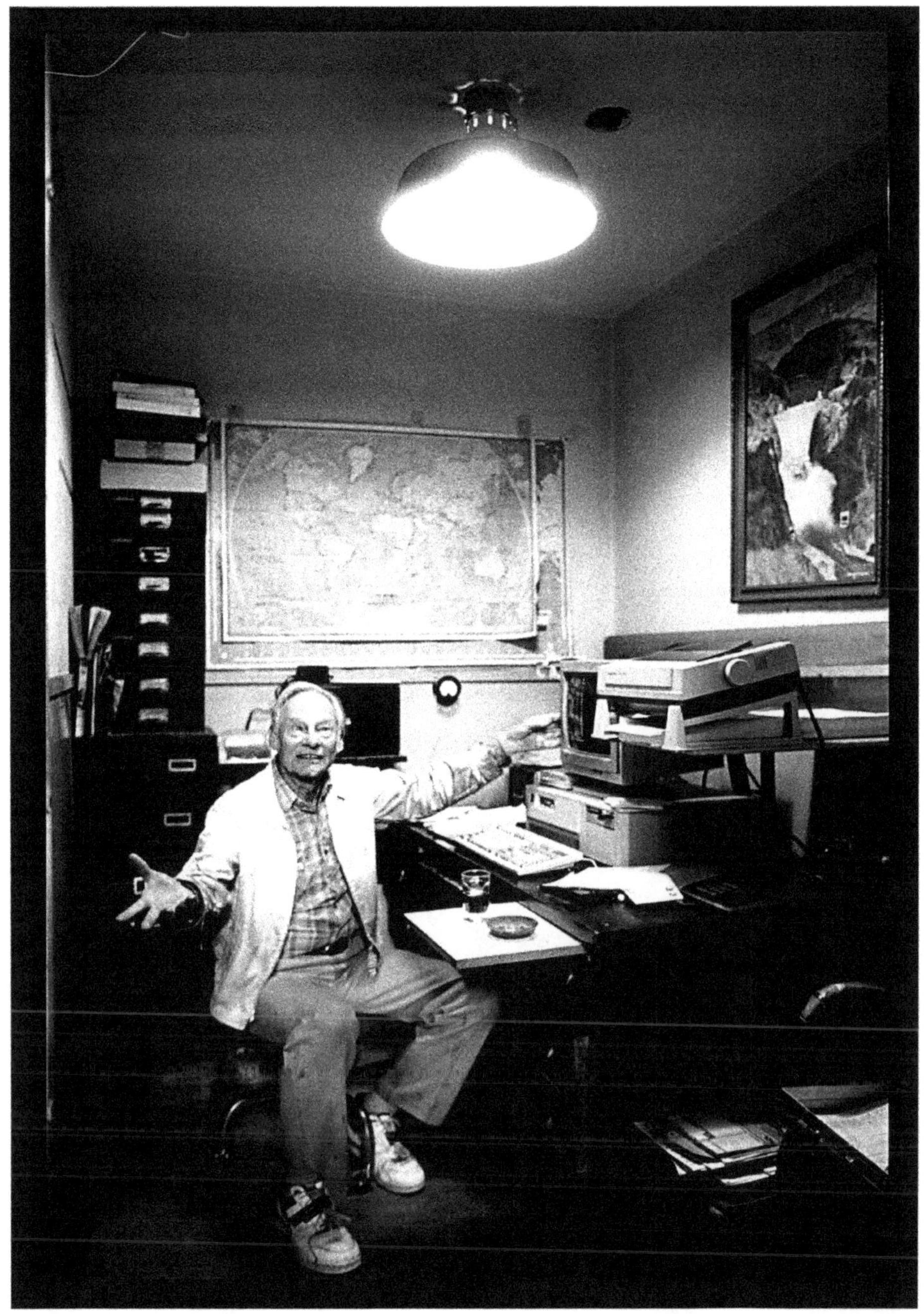

Alternative Housing. This Missile silo was constructed in 1960 at a cost of $5.5 million. It was purchased by engineer Francis Dellenbach for $3,011. He has converted the installation with attached command center and robust air conditioning into a home, office, and vehicle storage center. *Chugwater, Wyoming, 1988.*

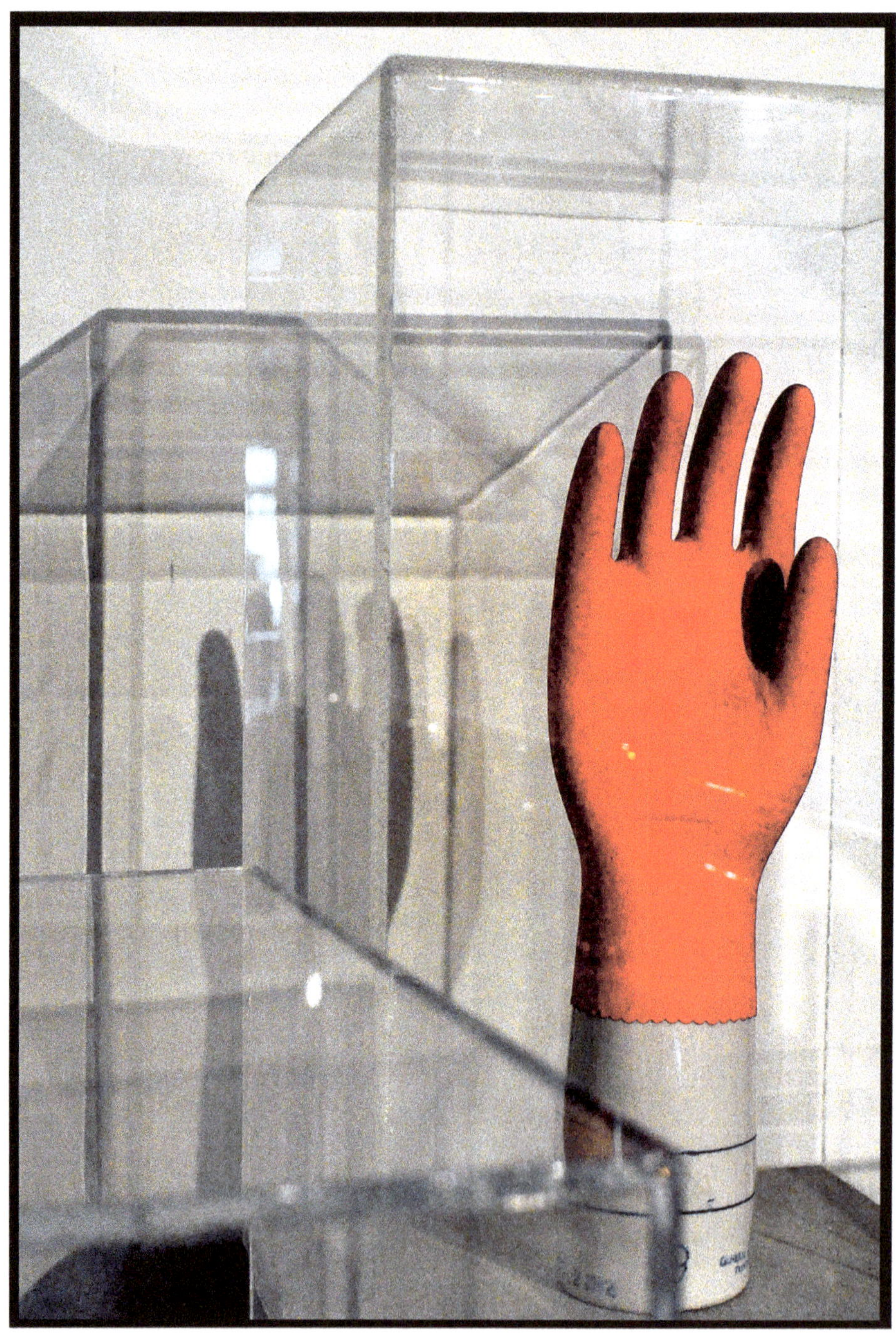

Protective Glove allows nuclear workers to safely handle contaminated materials. *National Museum of Nuclear Science and History, Albuquerque, New Mexico, 2013.*

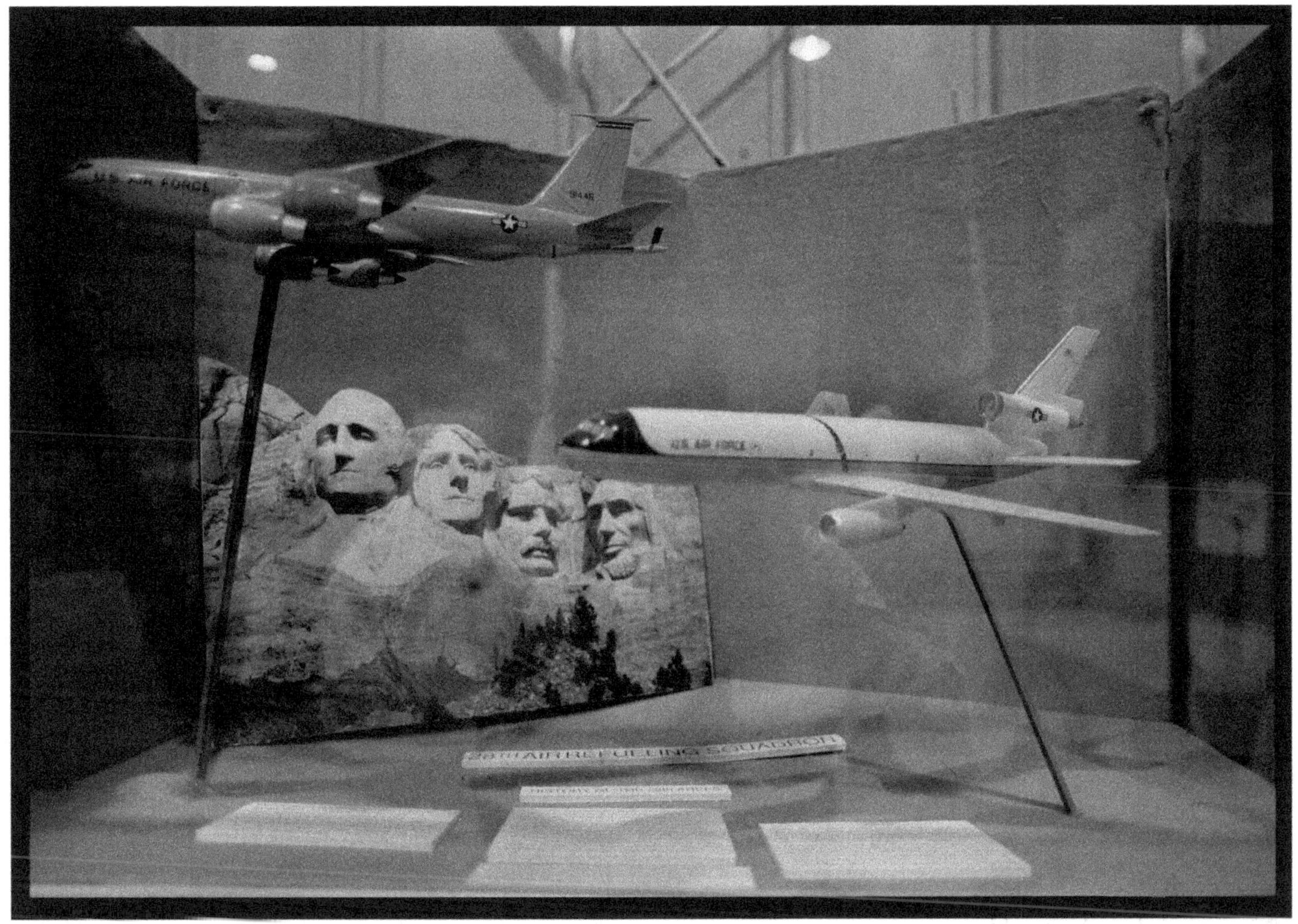

When Deterrence Fails the 28th Refueling Squadron springs into action. Based out of Ellsworth Air Force Base near Mount Rushmore, the 28th will refuel strategic bombers in midair so they can remain a viable threat for longer times over greater distances to deliver weapons to targets or engage in surveillance at high altitudes. *South Dakota Air and Space Museum, Ellsworth Air Force Base, Rapid City, South Dakota, 1989.*

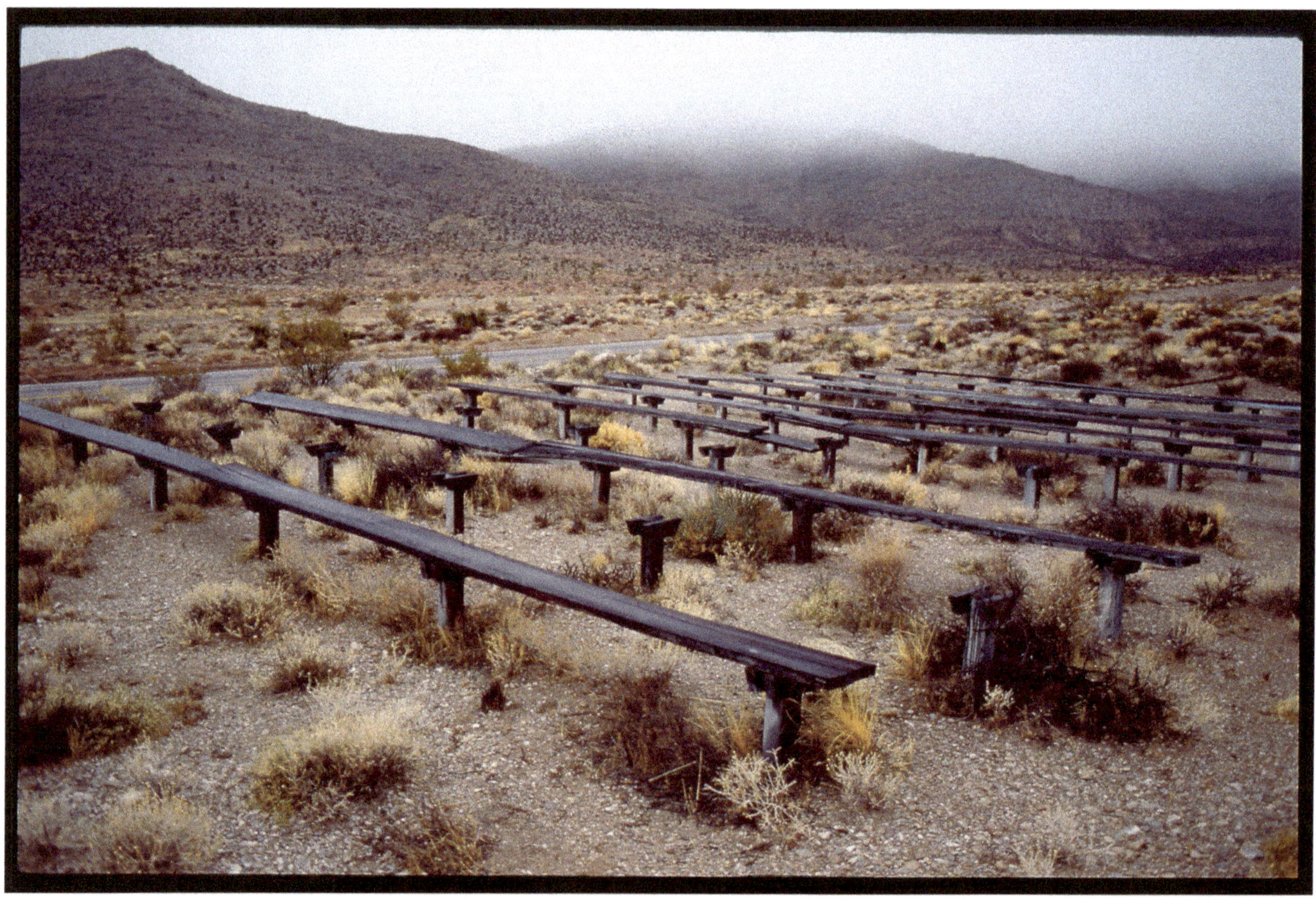

Abandoned Press Bleachers once accommodated scores of U.S. and international journalists eager to witness live atmospheric nuclear explosions. Over 100 above-ground blasts happened here between 1951 and 1962. Then testing continued underground. A total of 1,021 nuclear detonations took place within this testing range. *News Nob, Yucca Flats, Nevada Test Site, Nye County, Nevada, 1989.*

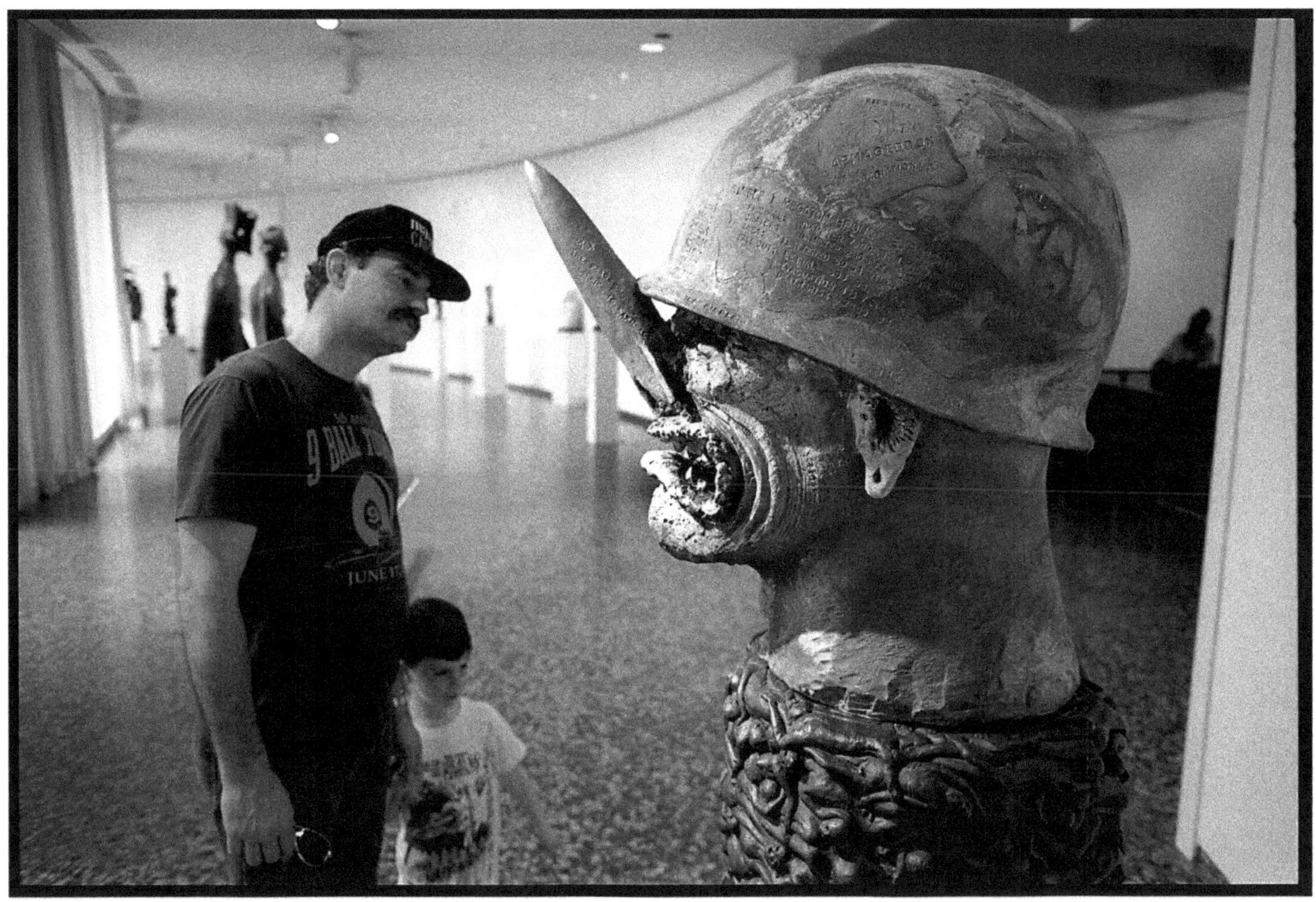

"General Nuke" is sculptor Robert Arneson's response to the warlike mood in the early 1980s when the U.S. challenged the U.S.S.R. by aiming intermediate-range European-based Pershing missiles at Russia. Wearing a 3-star General's helmet, with an MX intercontinental ballistic missile for a nose, General Nuke's hideous visage touches a dire nerve in all who look upon him. Sculpture by Robert Arneson, glazed ceramic and bronze on granite base, 1984. *Hirschorn Museum and Sculpture Garden, Washington, D.C., 2005.*

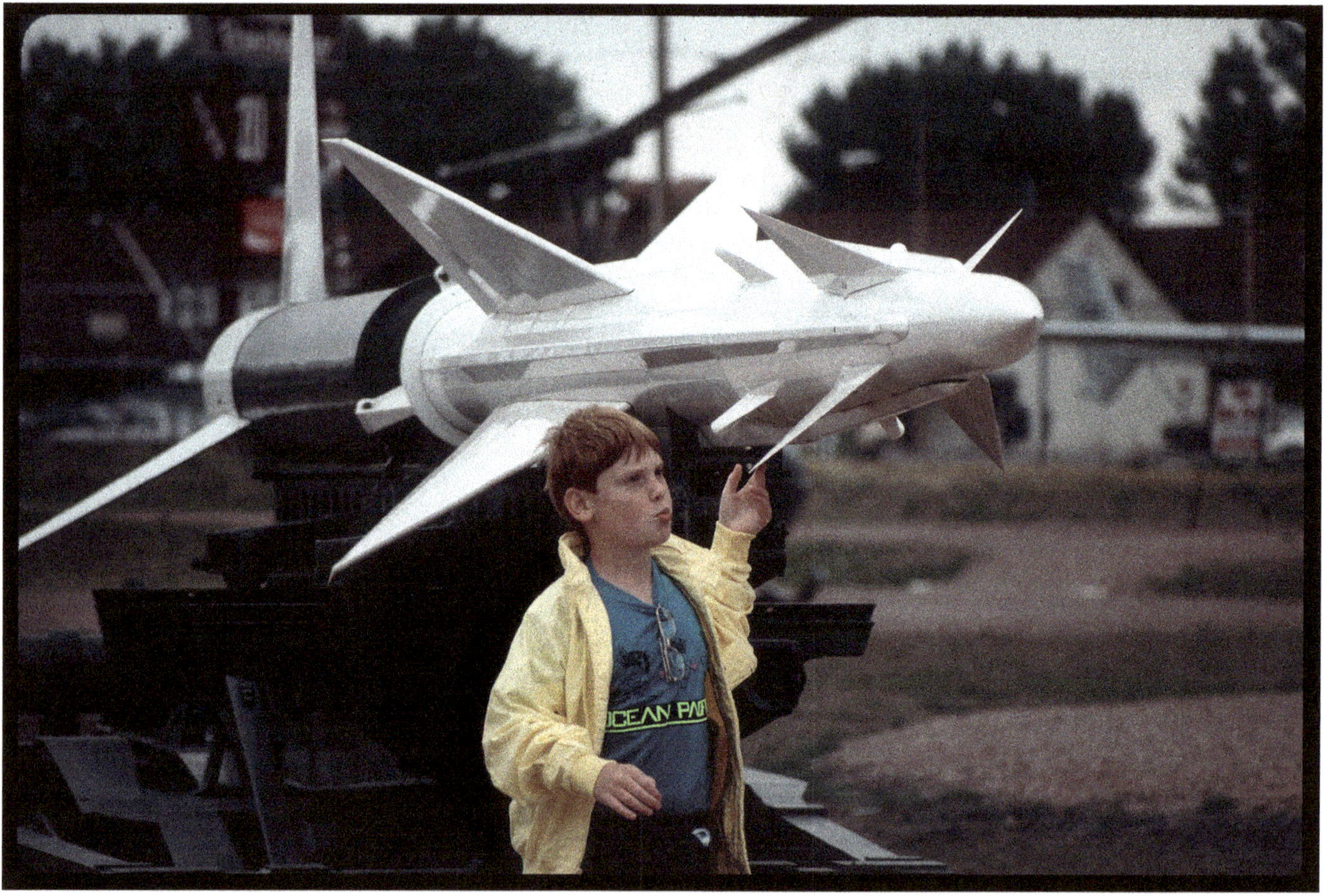

Imagining the Future. A young American, proud of his country's deterrent posture, fondles the forward fin of a tactical nuclear missile while peering into the distance, confident that the U.S. nuclear arsenal keeps his country safe. *Ellsworth Air Force Base, Rapid City, South Dakota, 1989.*

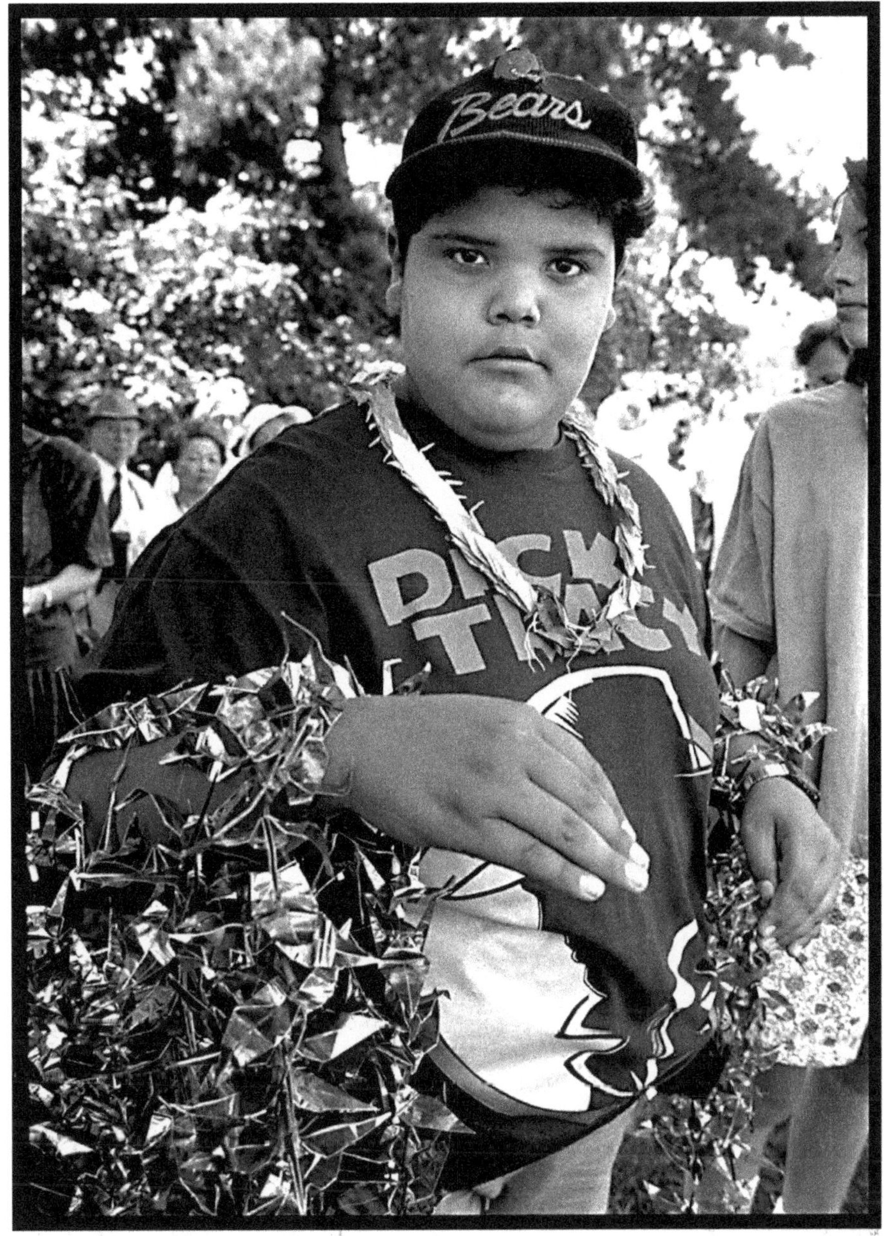

Remembering Hiroshima. A young American honors the victims of the Hiroshima atomic bomb by gathering around himself 1,000 paper cranes on the forty-second anniversary of the dropping of the Bomb. One thousand paper cranes have long been a symbol of Sadako Sasaki's hope-against-hope in the nuclear age. *Seattle, Washington, August 6, 1987.*

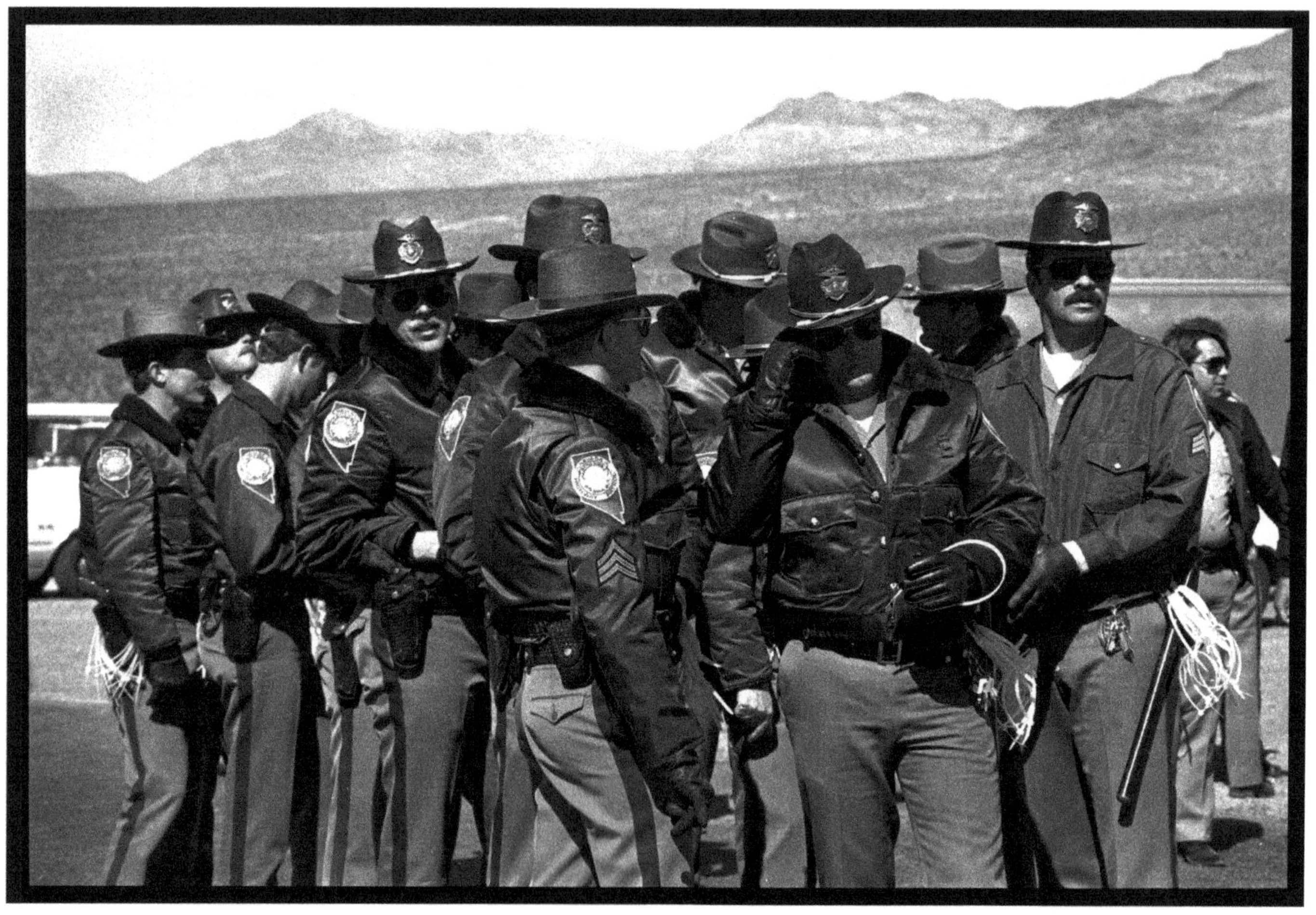

Nevada Test Site Sheriffs wait to arrest, handcuff, and detain protestors who cross the test site property line during one of the many protests held here. *Nevada Test Site, Nye County, Nevada, March 12, 1988.*

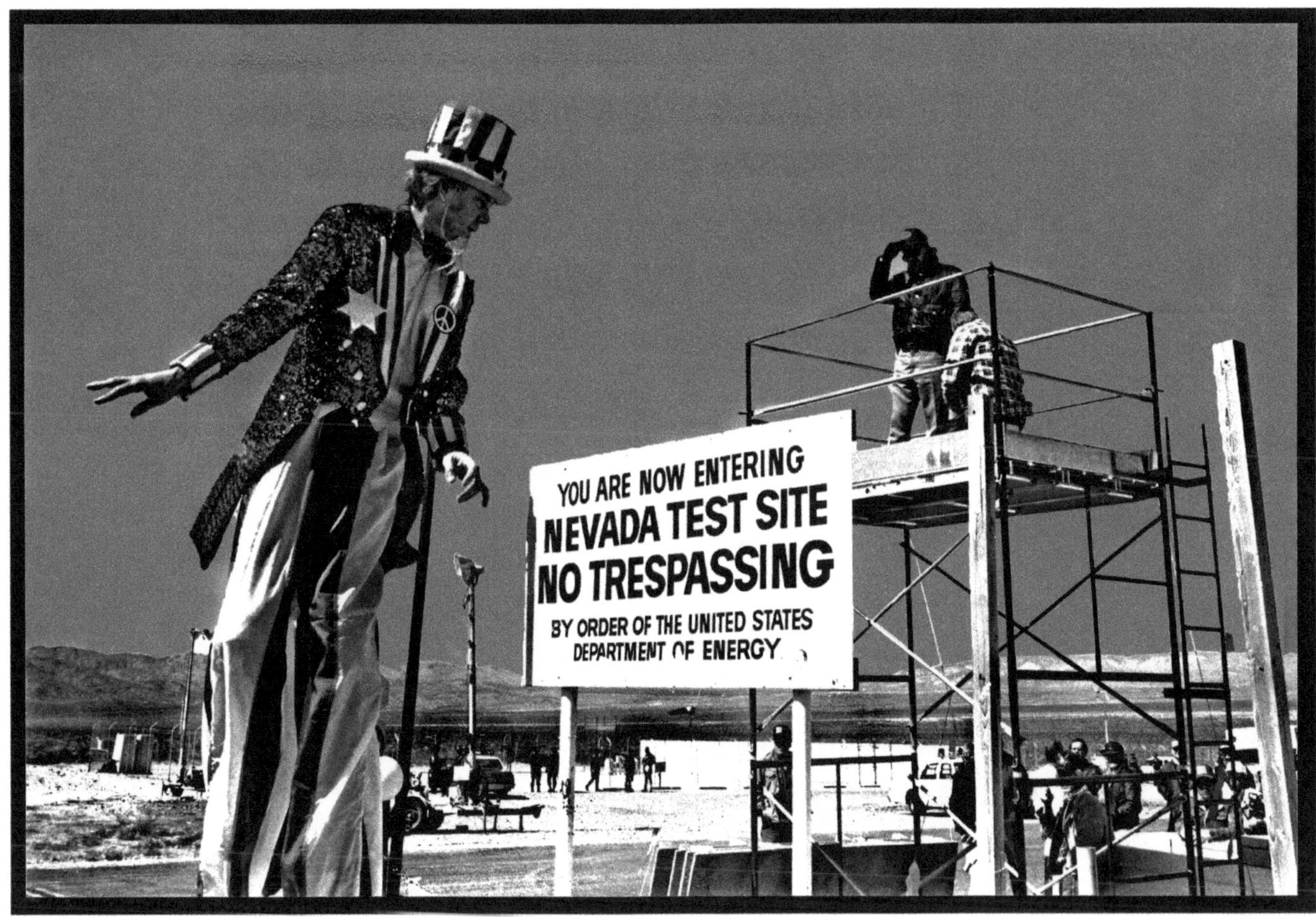

Walking the Line. Uncle Sam hops, kicks, and dances close to the Nevada Test Site boundary. Guards primed to cuff and detain watch and wait. *Nevada Test Site, Nye County, Nevada, March 12, 1988.*

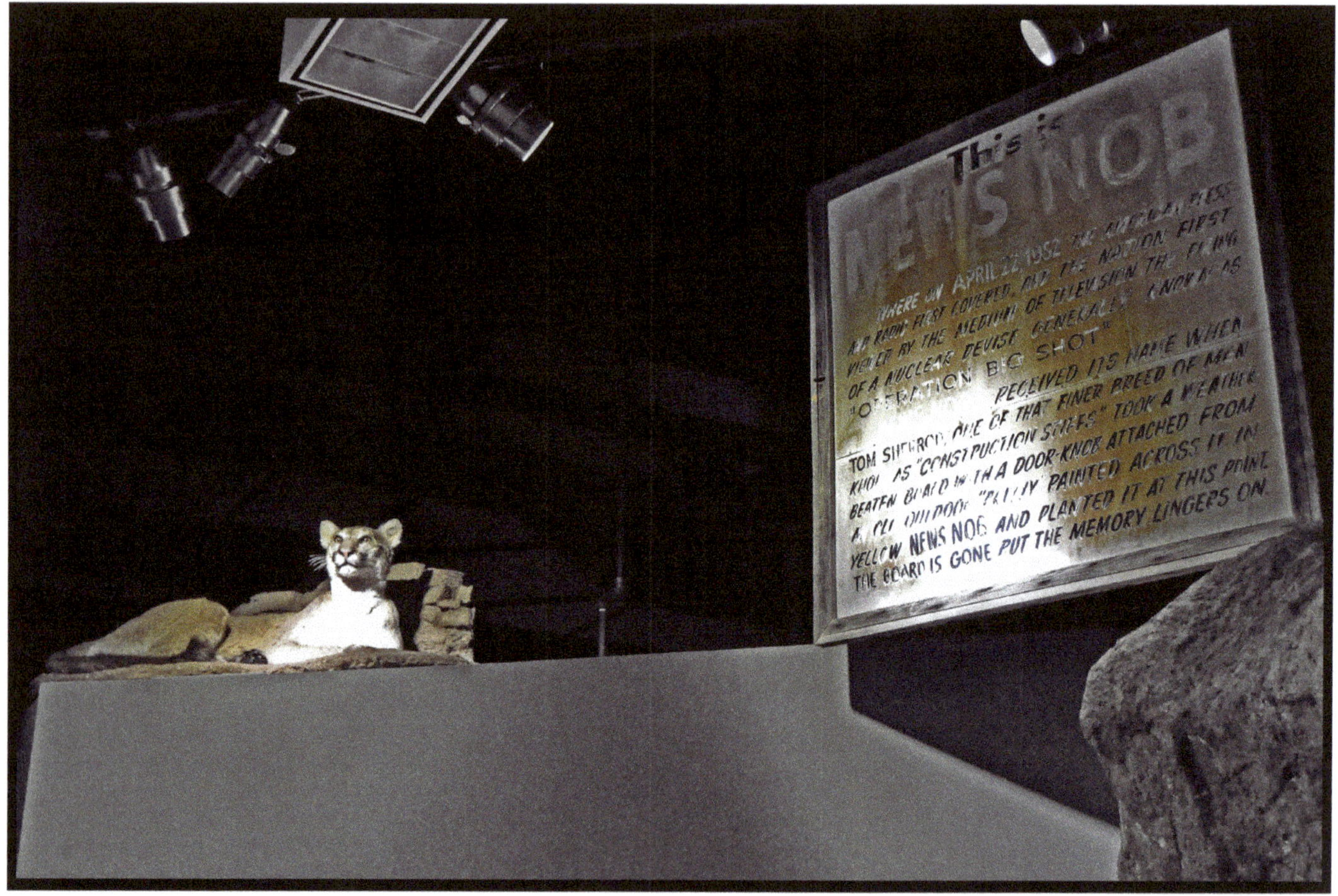

The Text on this Sign Reads, “This is News Nob Where on April 22, 1952, the American press and radio first covered, and the nation first viewed by the medium of television—the firing of a nuclear device generally known as ‘Operation Big Shot.’ News Nob got its name when Tom Sherrod, one of that finer breed of men known as ‘Construction Stiffs’ took a weather beaten board with a doorknob attached from an old outdoor ‘privy,’ painted across it in yellow ‘News Nob,’ and planted it at this point. The board is gone but the memory lingers on.” *Atomic Testing Museum, Las Vegas, Nevada, 2013.*

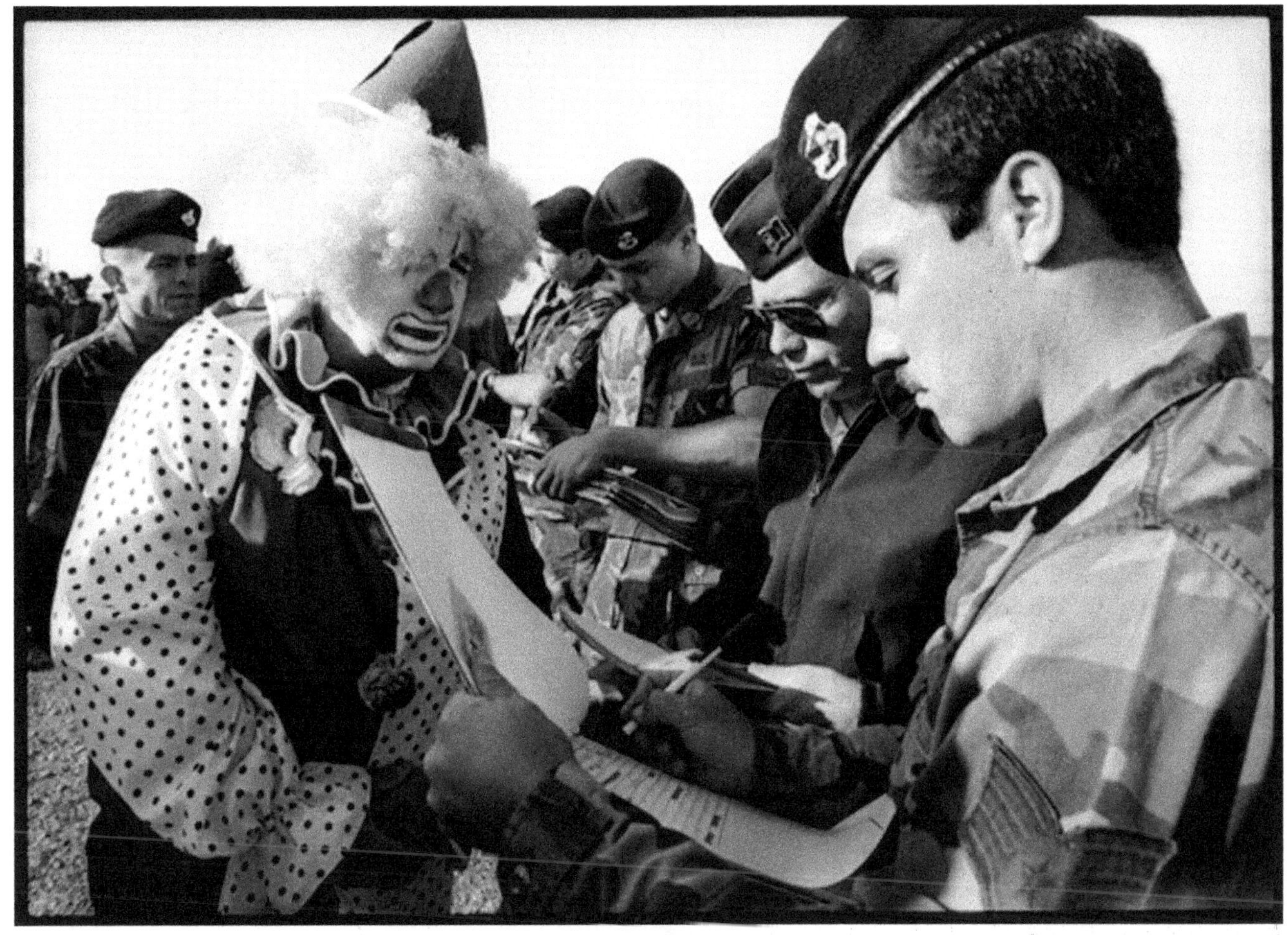

Sam Day climbed an active missile silo for which he was arrested. Sam Day is the author of *Nuclear Heartland*; it lists the location of every missile silo site in the United States. *Knob Noster, Missouri, 1991.*

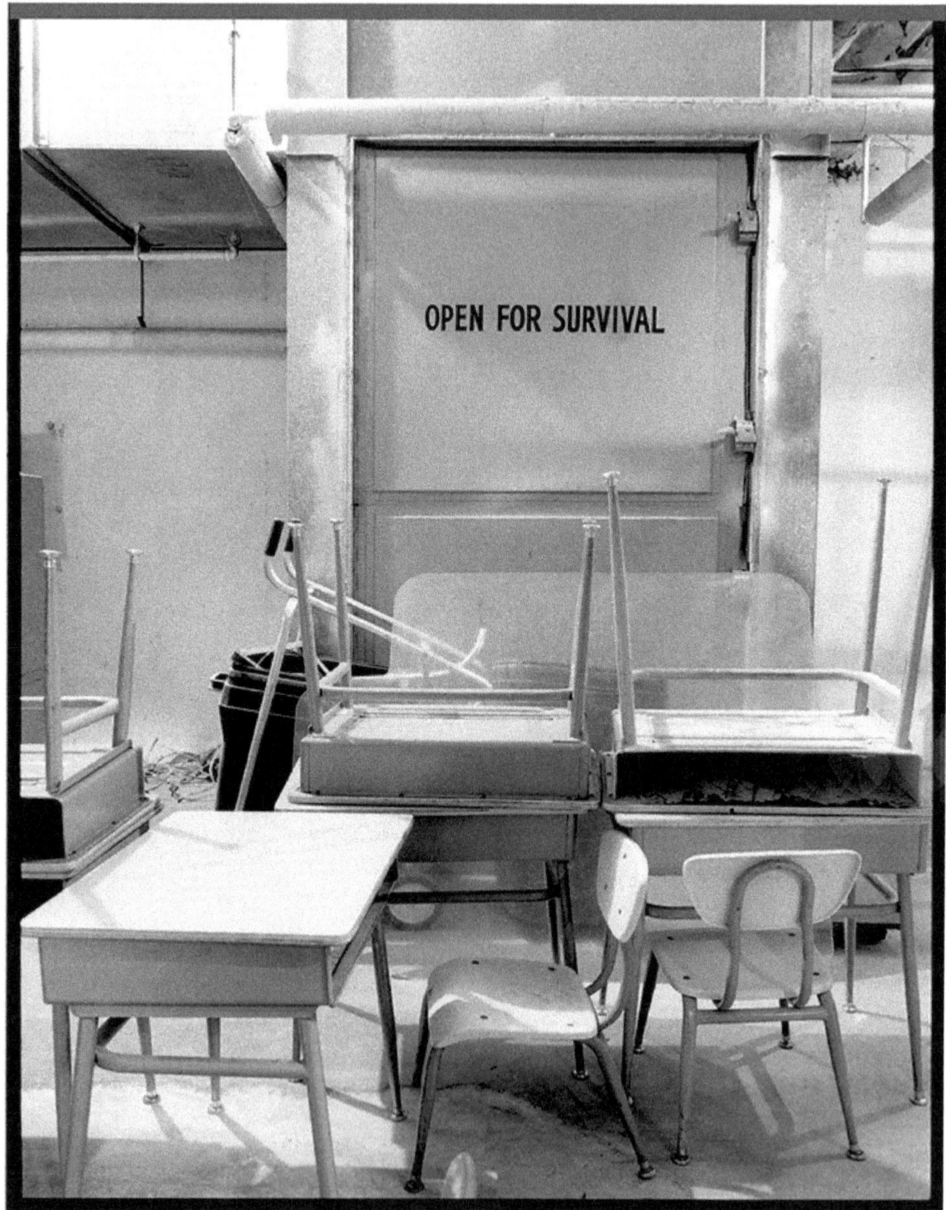

Duck and Cover. In the era when children were taught that hunkering down under their desks at school would save them from the Bomb, the ABO Elementary School—home of the Gophers—became, in 1962, the nation's first underground public school bomb shelter. In a community of 12,000, the shelter was able to accommodate 2,160 people and a morgue. Students spent long days in windowless rooms. How they adapted was studied. President Kennedy congratulated the school, and it received a federal engineering award for innovation. The AAA lists this underground school as a New Mexico tourist attraction. *ABO Elementary School and Fallout Shelter, Artesia, New Mexico, 1989.*

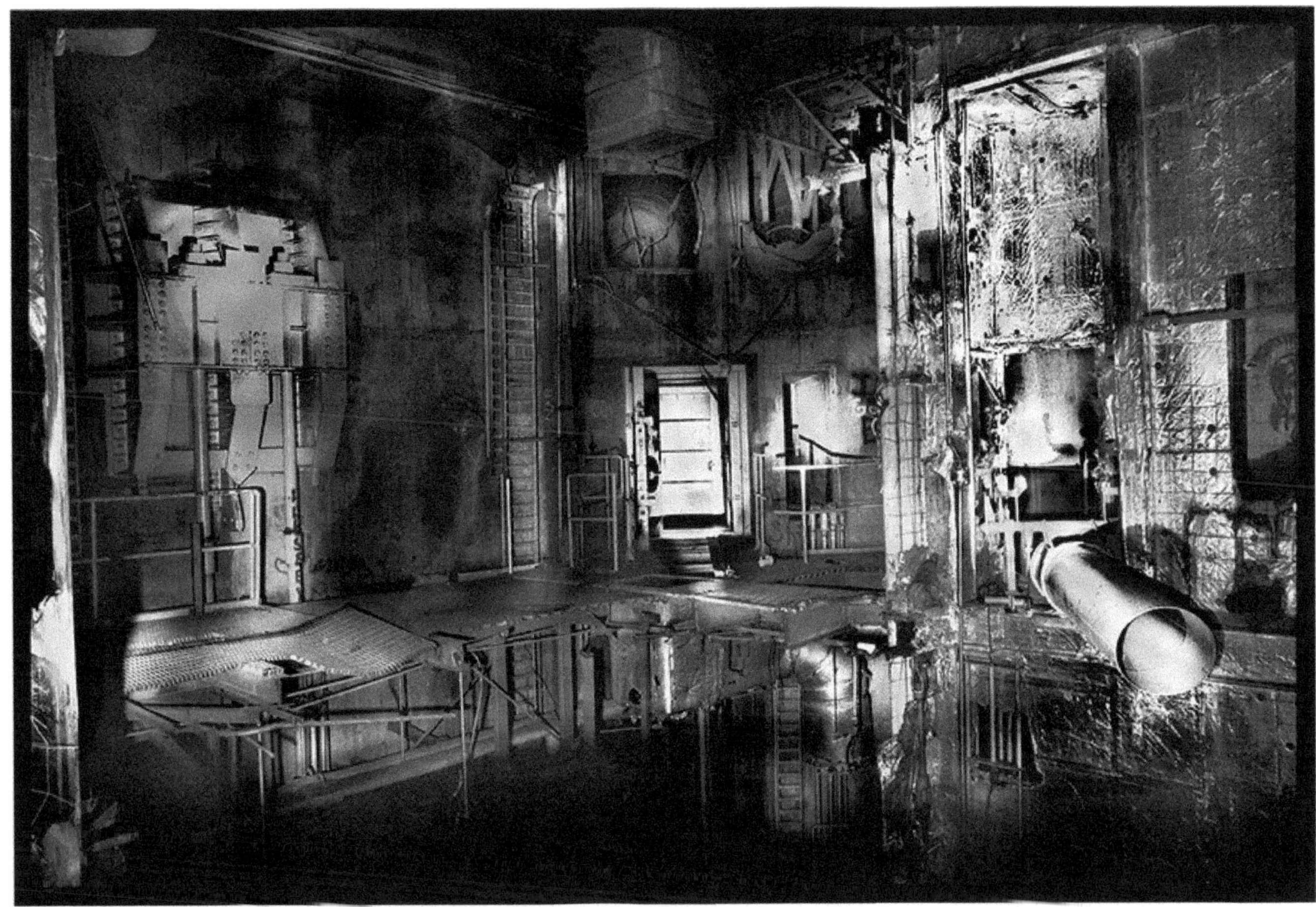

Deterrence Unhinged. This abandoned missile silo has been broken into with a blow-torch, stripped of copper wiring, relieved of marketable metals, sprayed with graffiti, its blast doors unhinged, and its safety barriers stolen—so that where once an Intercontinental Ballistic Atlas Missile stood on hair-trigger alert, now there is an eighty-five foot drop into darkness. *Atlas Missile Silo, Roswell, New Mexico, 1995.*

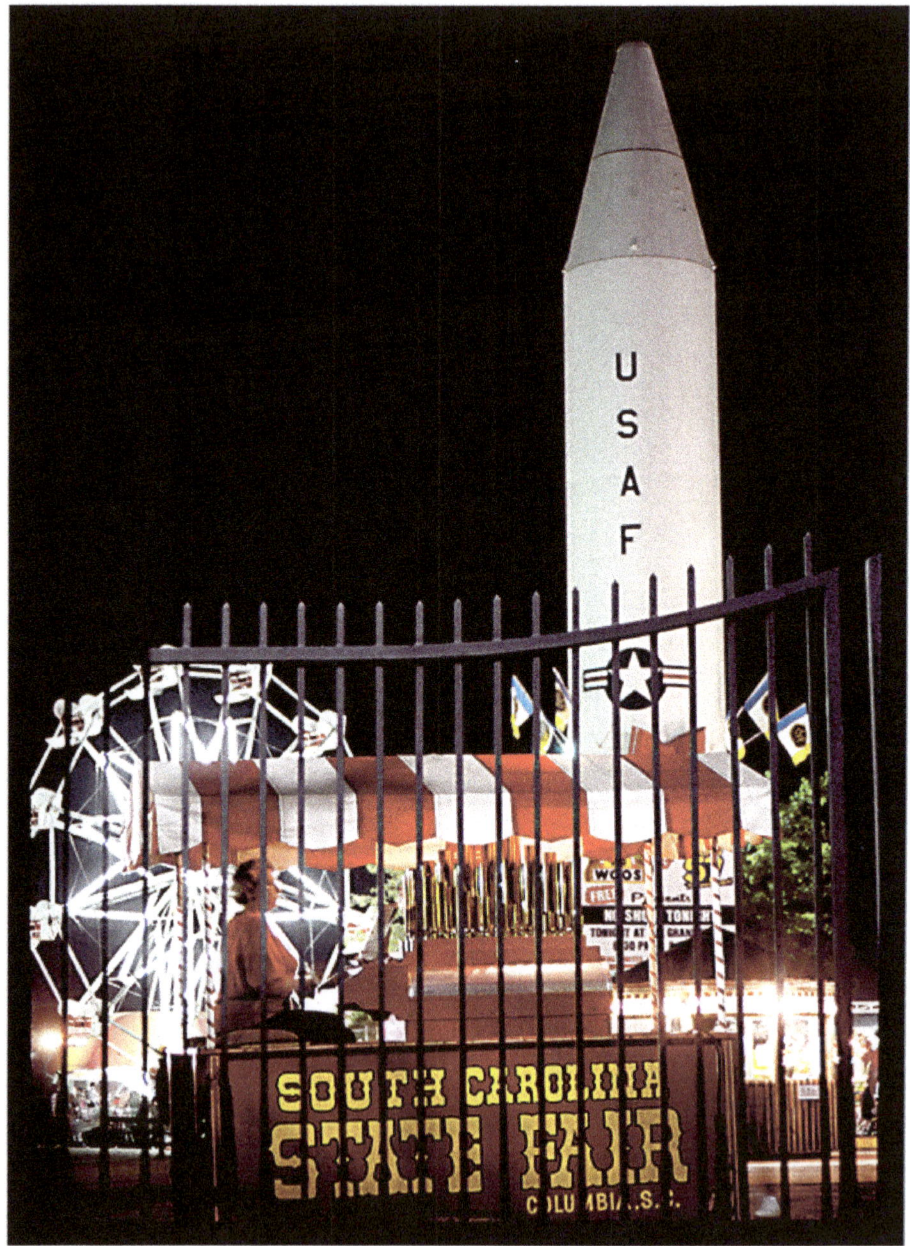

State Fair Missile. Instantly recognizable and the tallest thing in sight, this ICBM missile casing served as a meeting point for lost children and worried parents at the 1987 South Carolina State Fair. *Colombia, South Carolina, 1987.*

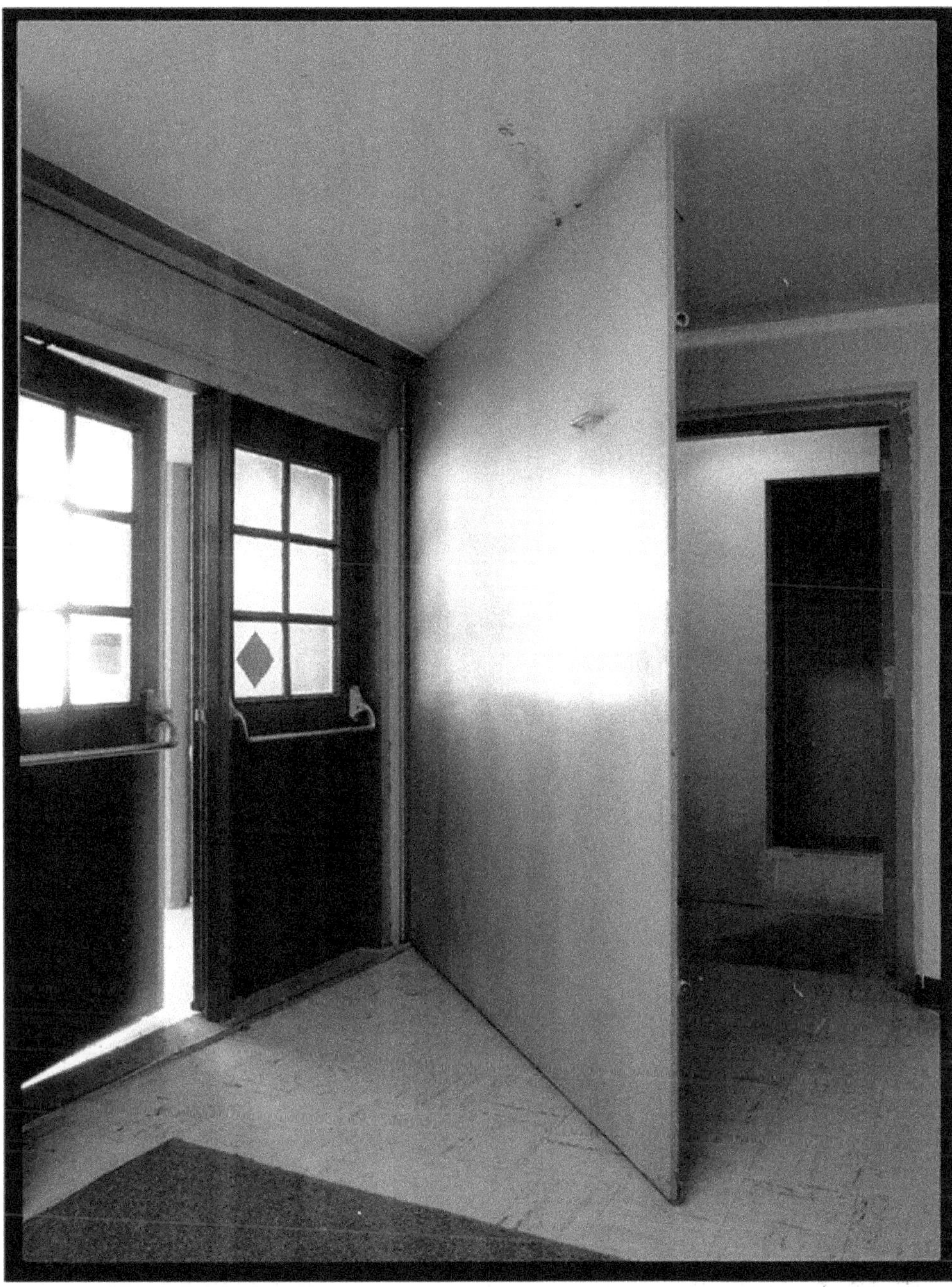

Steel Salvation. This portal, weighing in at 1800 pounds, stands ready to allow access to citizens fleeing fireballs and fallout. After 2,160 of the saved have crossed the threshold into the shelter, this door and two others like it get bolted shut as atomic dramas of destruction, salvation, and remorse play out in this village of 20,000 souls. *Underground ABO Elementary School, Artesia, New Mexico, 1989.*

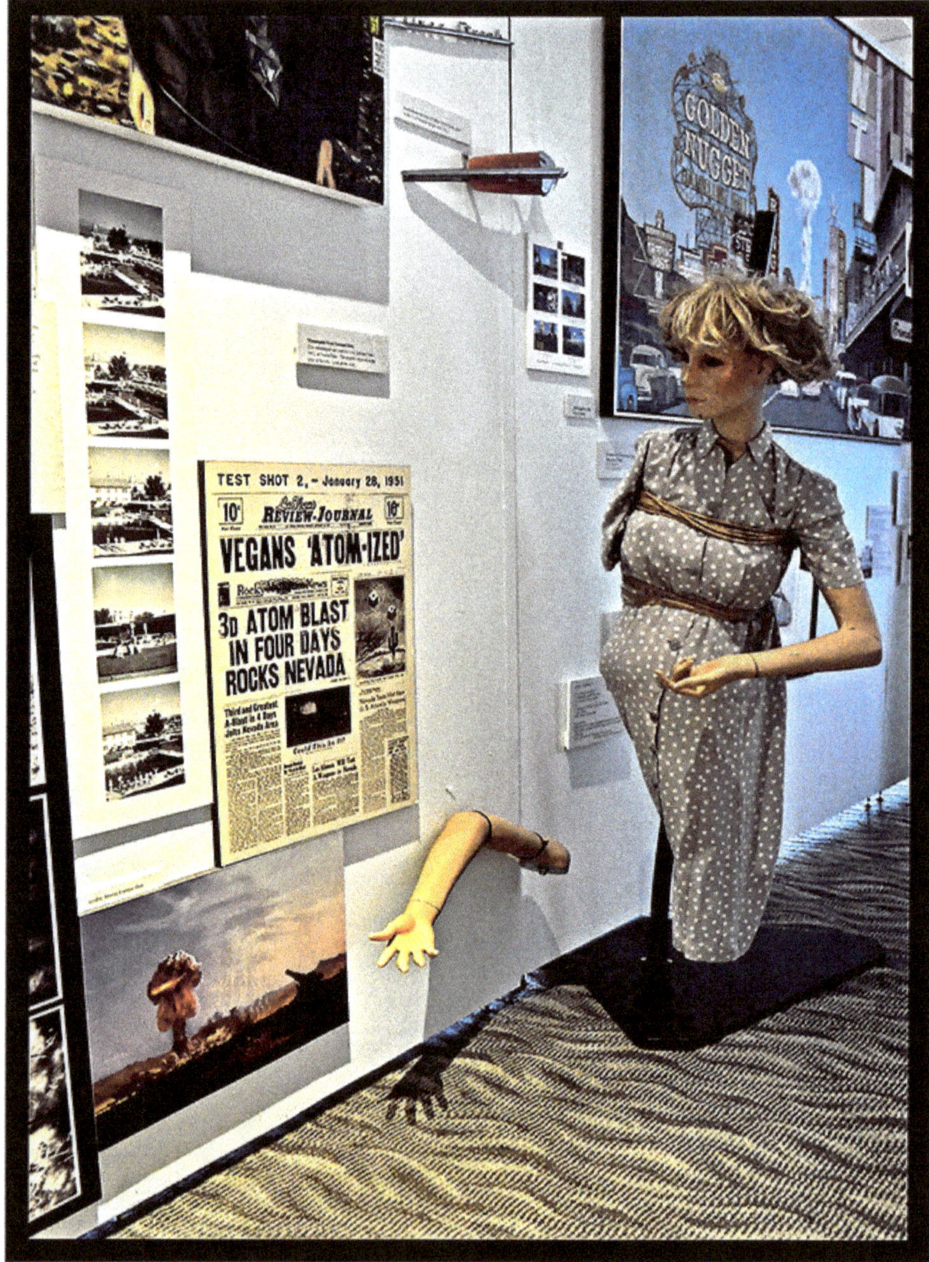

Armless in Vegas. A mannequin portraying a housewife dismembered by a nuclear blast stands in her JC Penny dress at a display entitled Building Atomic Vegas. The show explores the influence of nuclear testing on the life and culture of the city. At the test site itself, a life-size village called "Doom Town" was constructed close to a future ground zero. Las Vegas merchants outfitted these properties slated for atomic annihilation. *Atomic Testing Museum, Las Vegas, Nevada, 2011.*

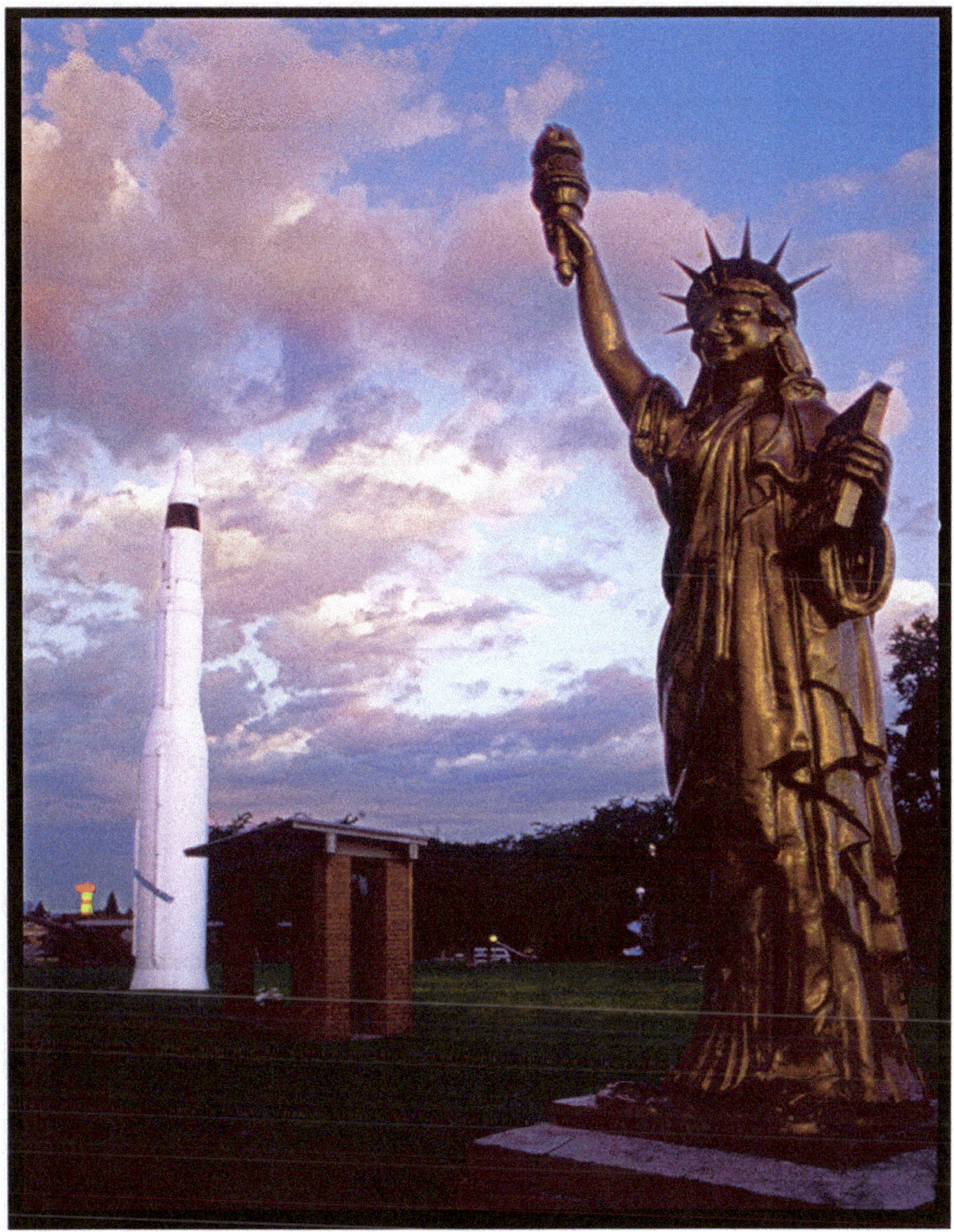

Minuteman and Lady Liberty. How many Americans does it take to put a plutonium warhead inside a Minuteman missile? From 1945 to 1986, when this photo was taken, 60,000 plutonium warheads for 116 weapons systems have been produced in factories in 13 states employing 90,000 workers for $230 billion (in 1986 dollars) with another $1.85 trillion (1986 dollars) spent on nuclear delivery systems. *Nuclear Weapons Data Book*, vol. 2, by Cochran, Arkin, Norris, and Hoening, Natural Resources Defense Council (Ballinger). *Lewistown, Montana, 1988.*

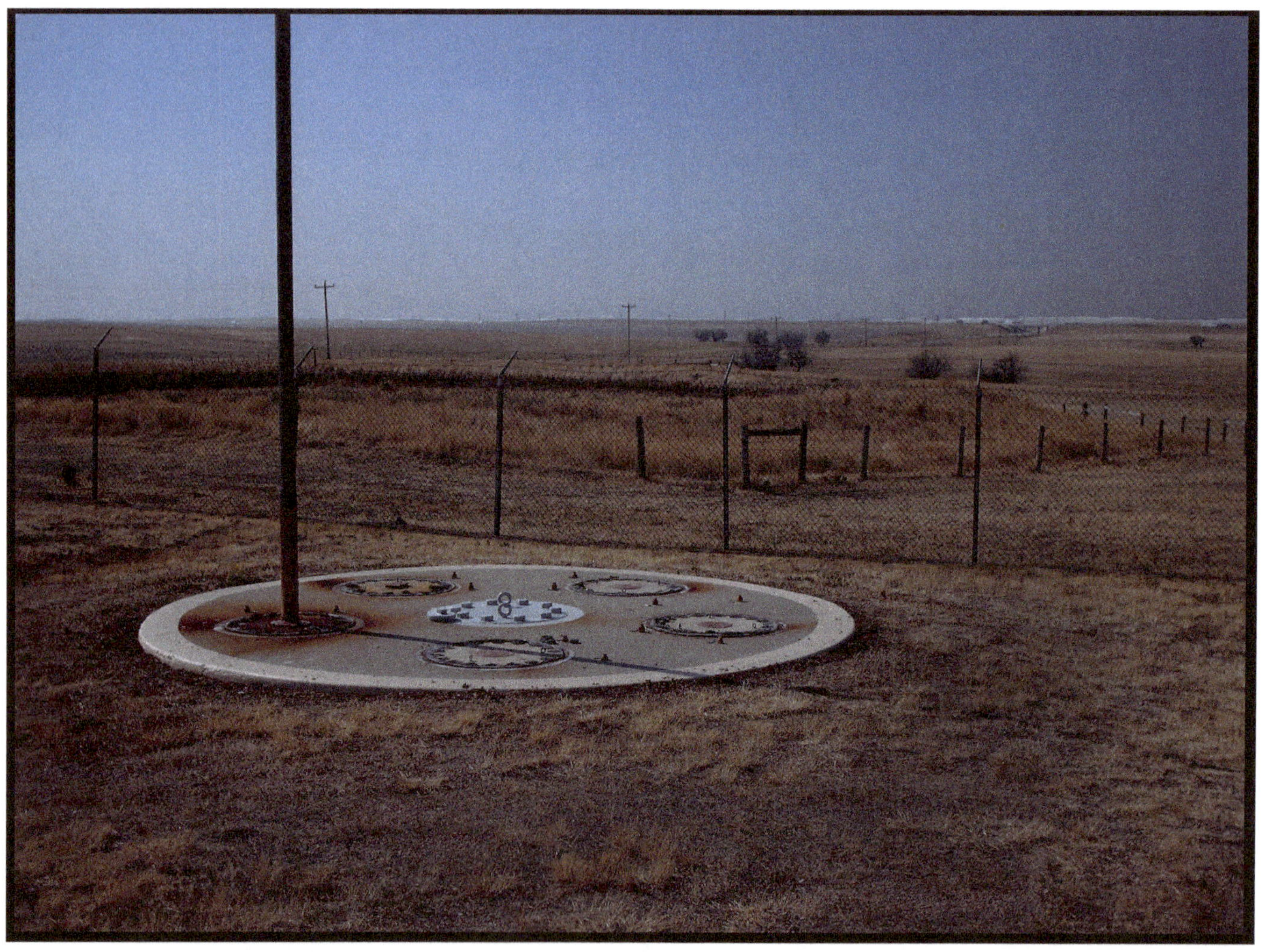

This High Frequency Antenna was designed to track the trajectory of an ICBM to its final destination.
Minuteman Missile Silo, Phillip, South Dakota, 1988.

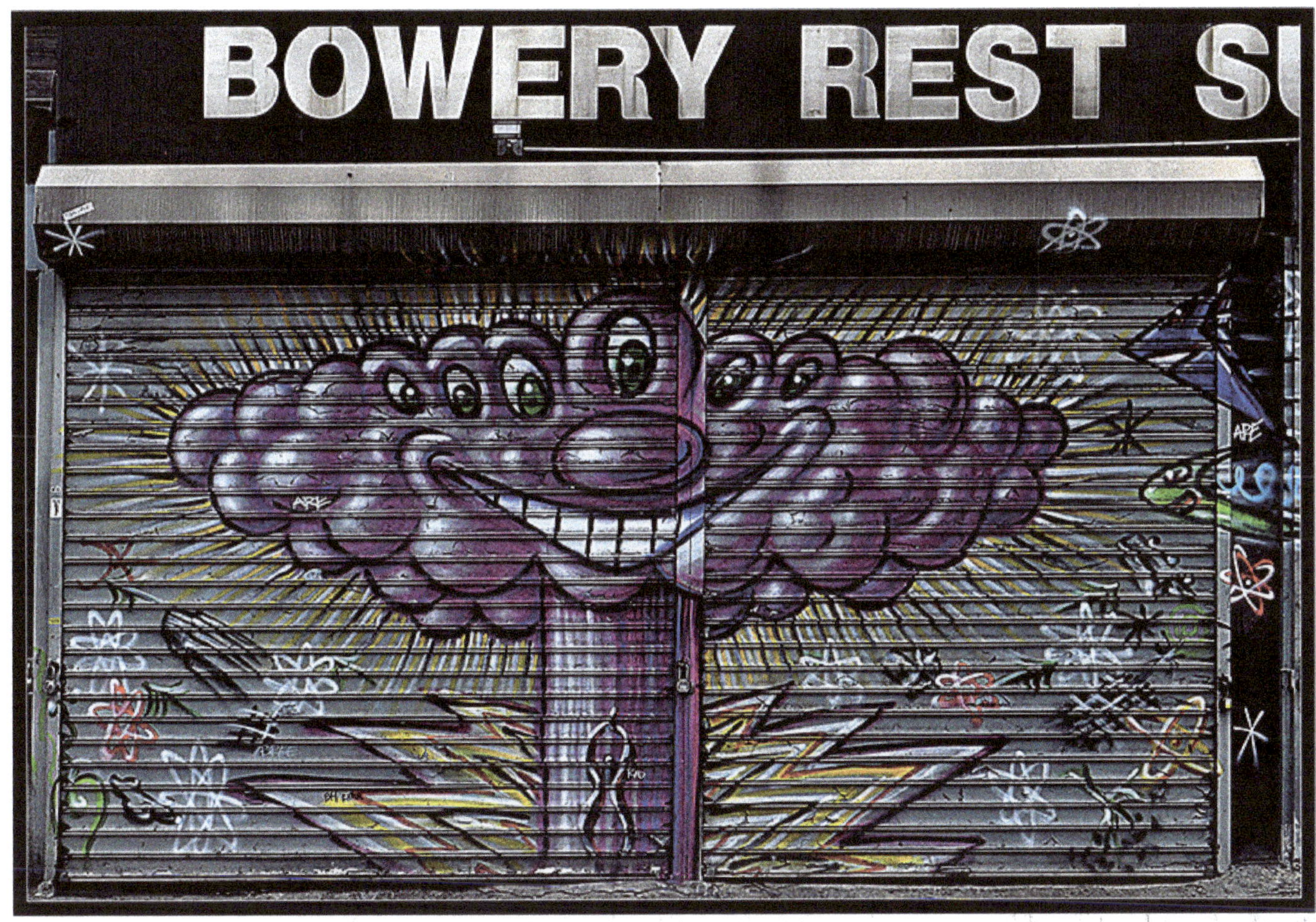

A Mushroom Cloud with Attitude graces the front of a small business establishment in Manhattan. It was created by graffiti artist Kenny Scharf, who specializes in painting ballooney figures and atomic clouds throughout New York City. *New York, 2012.*

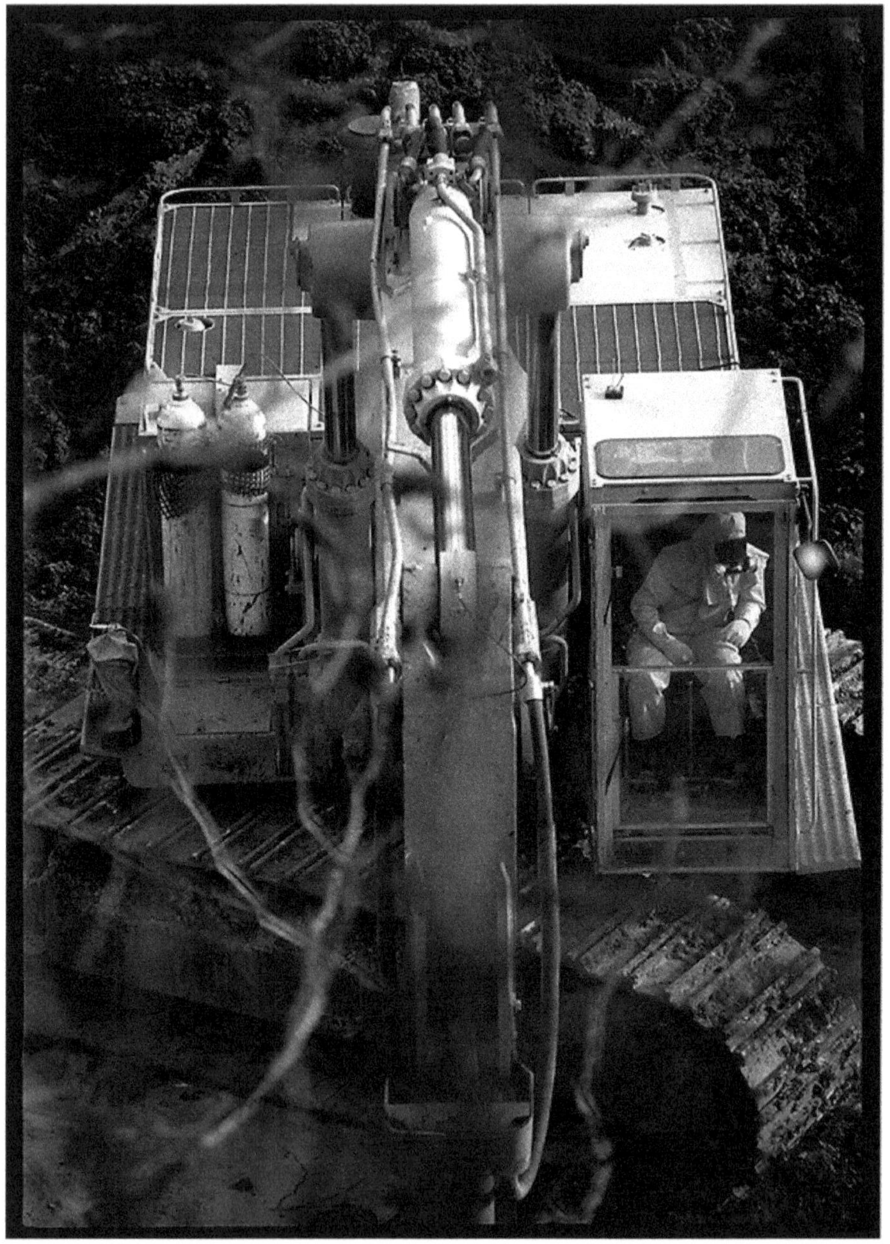

Hydraulic Excavator at Weldon Spring helped move 1.5 million cubic yards of uranium-contaminated wastes on the site of the Weldon Spring uranium refinery. The wastes were deposited into a 45 acre disposal mound at the same location. The process was mandated by the Comprehensive Environmental Response, Compensation, and Liability Act (CERCLA). It lasted 15 years, from 1986 until 2001. *Weldon Spring, Missouri, 1995.*

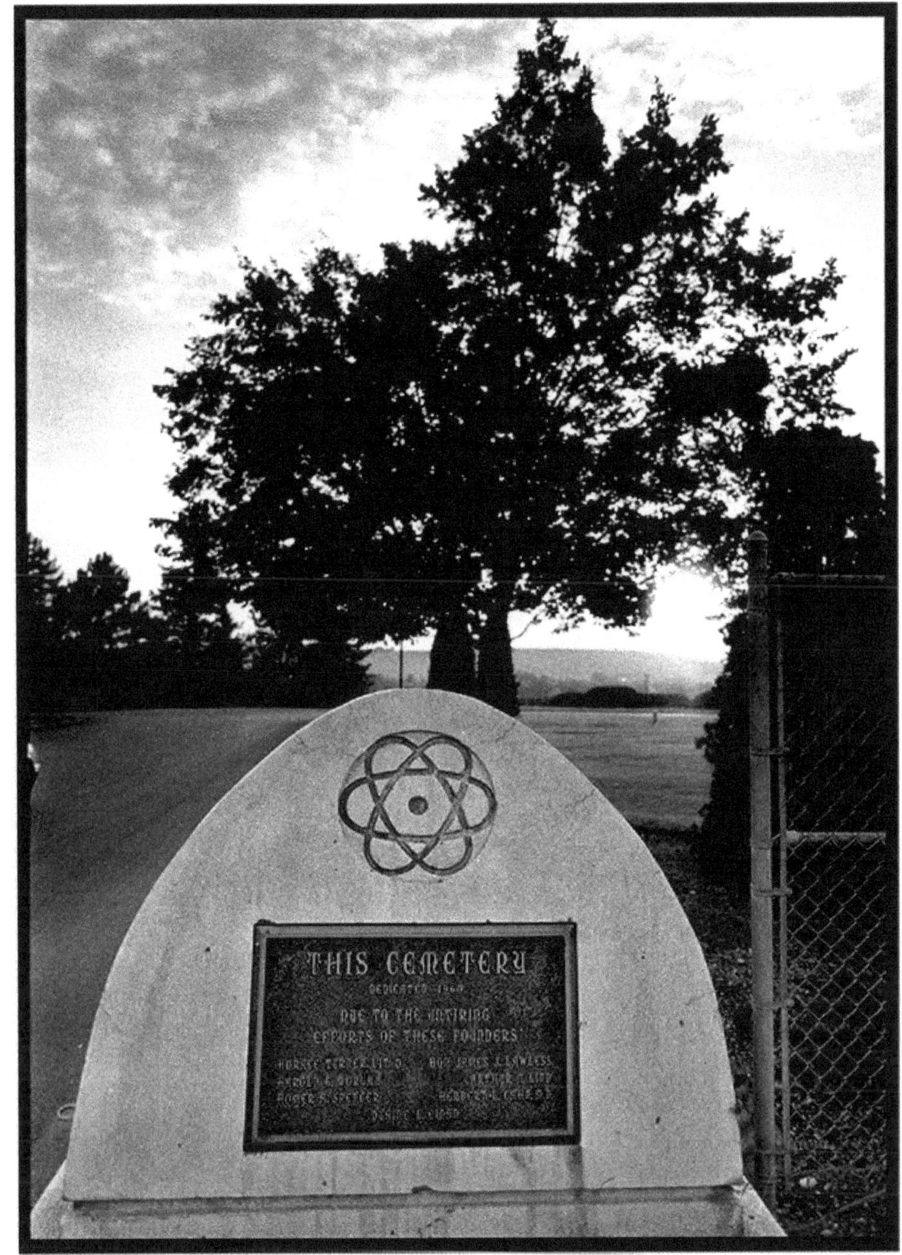

Nuclear Headstone identifies the final resting place for many employees from the Hanford Nuclear Reservation. It serves as a reminder to all who visit this ground, or who plan to be interred within it, that Hanford workers' lives were spent making America safe by helping create plutonium for over 60,000 nuclear warheads. *Richland Cemetery, Washington, 1992.*

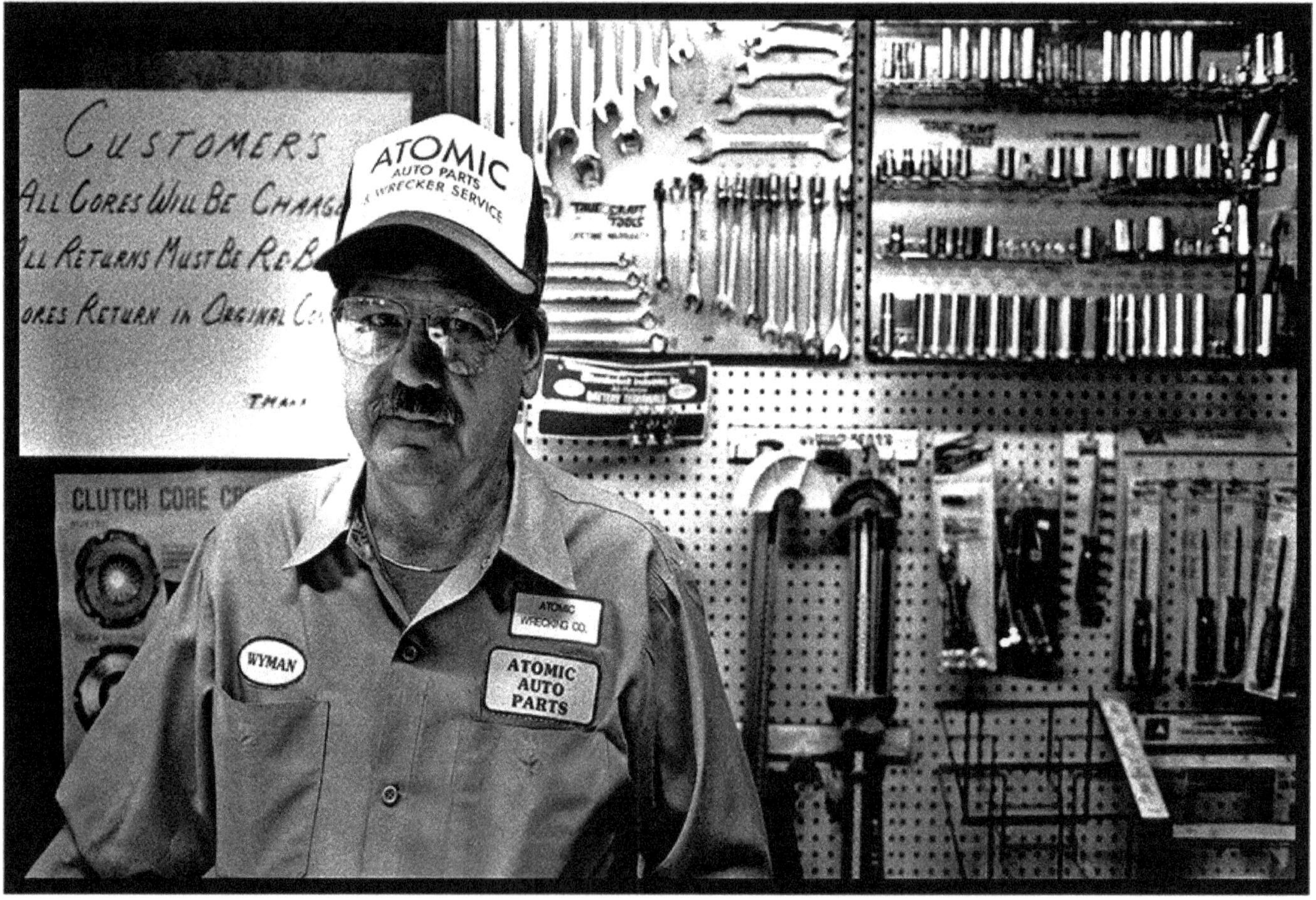

Tow Truck Operator near Atomic Road, a stone's throw from the Savanah River site's five plutonium production reactors. *Atomic Auto Parts and Wrecker Service, 3921 Whiskey Road, Aiken, South Carolina, 1989.*

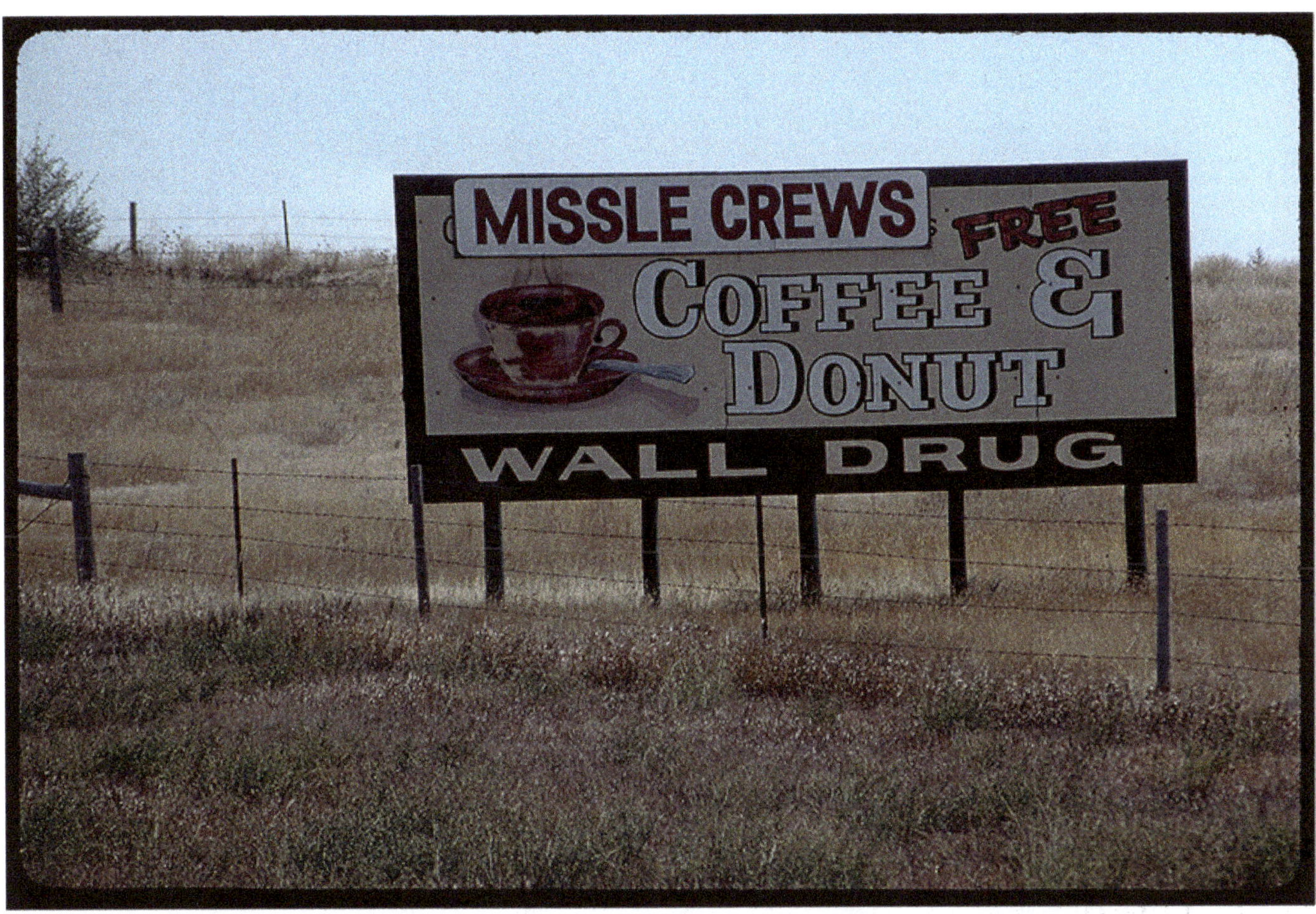

One More Cup of Coffee for the Road. Currently there are 450 land-based Minuteman III ICBMs in the US arsenal. Ten missiles are controlled by each Launch Control Center staffed by two key-turning missileers, underground, and, top-side, a commander, six security guards, and a chef cooking up three squares a day for each missileer. Back-up teams from regional air force bases provide fresh replacements. Minuteman missiles are exclusively stationed in four states: Colorado, North Dakota, Nebraska, and South Dakota. Wall Drug is a single store in a small South Dakota town, but its billboards and bumper stickers appear around the world. If the free coffee offer were made today, nearly 1,000 crew members could take advantage of it. Savings, however, are minimal: coffee at Wall Drug still sells for 5 cents a cup. *Billboard located on 1-90 near Wall, South Dakota (population 766), 2006.*

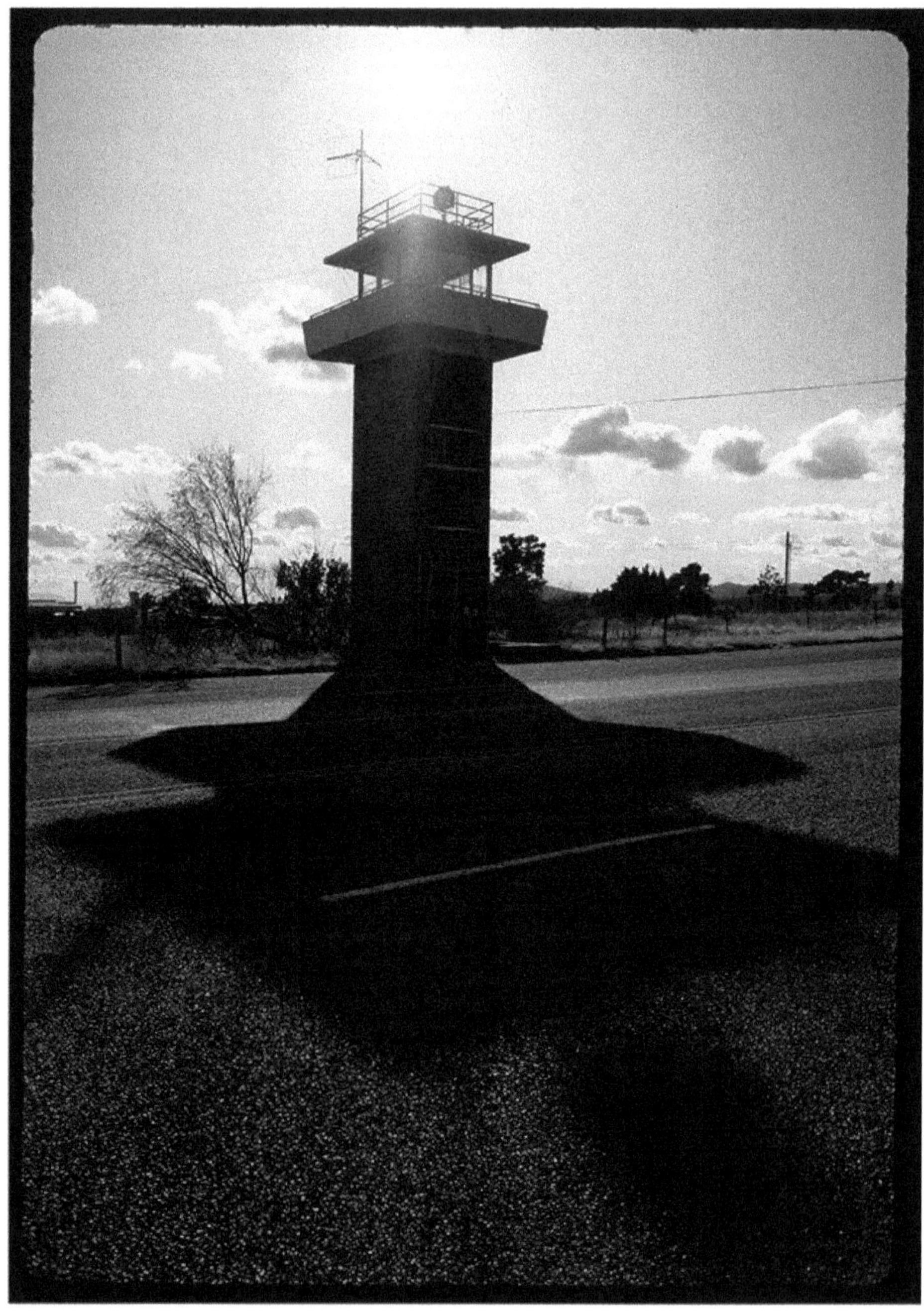

Cold War Brick Tower still hovers over the entrance to The Los Alamos National Laboratory, the birthplace of the atomic bomb. Los Alamos remains, with Lawrence Livermore, one of the two nuclear weapons design labs in the USA. *Los Alamos, New Mexico, 2008.*

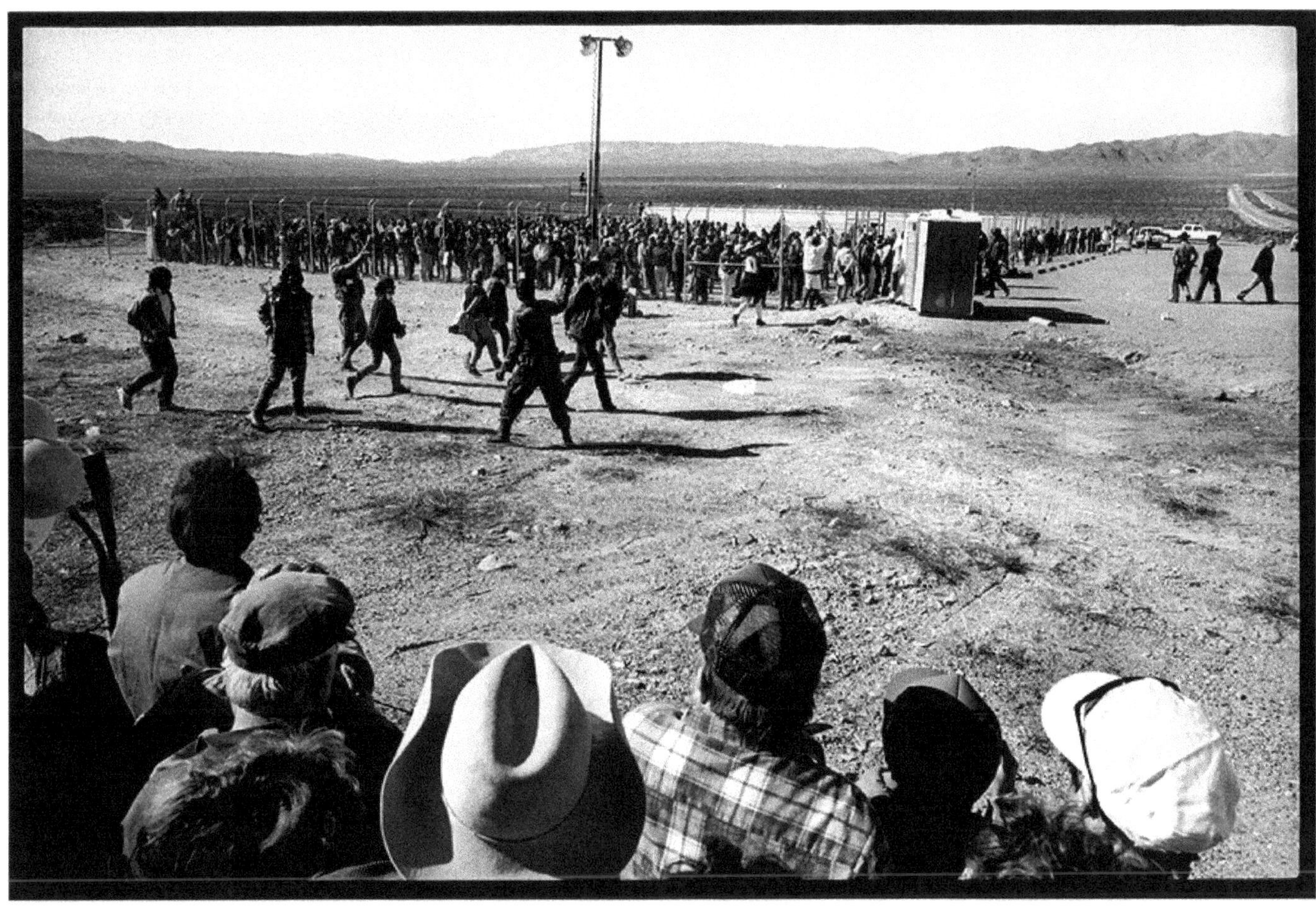

Protestors Outflank Sheriffs, climb over the perimeter fence in groups, walk to holding pens to be arrested, get bused to Tonopah 90 miles away, and are released to find their own way home. *Nevada Test Site, 1988.*

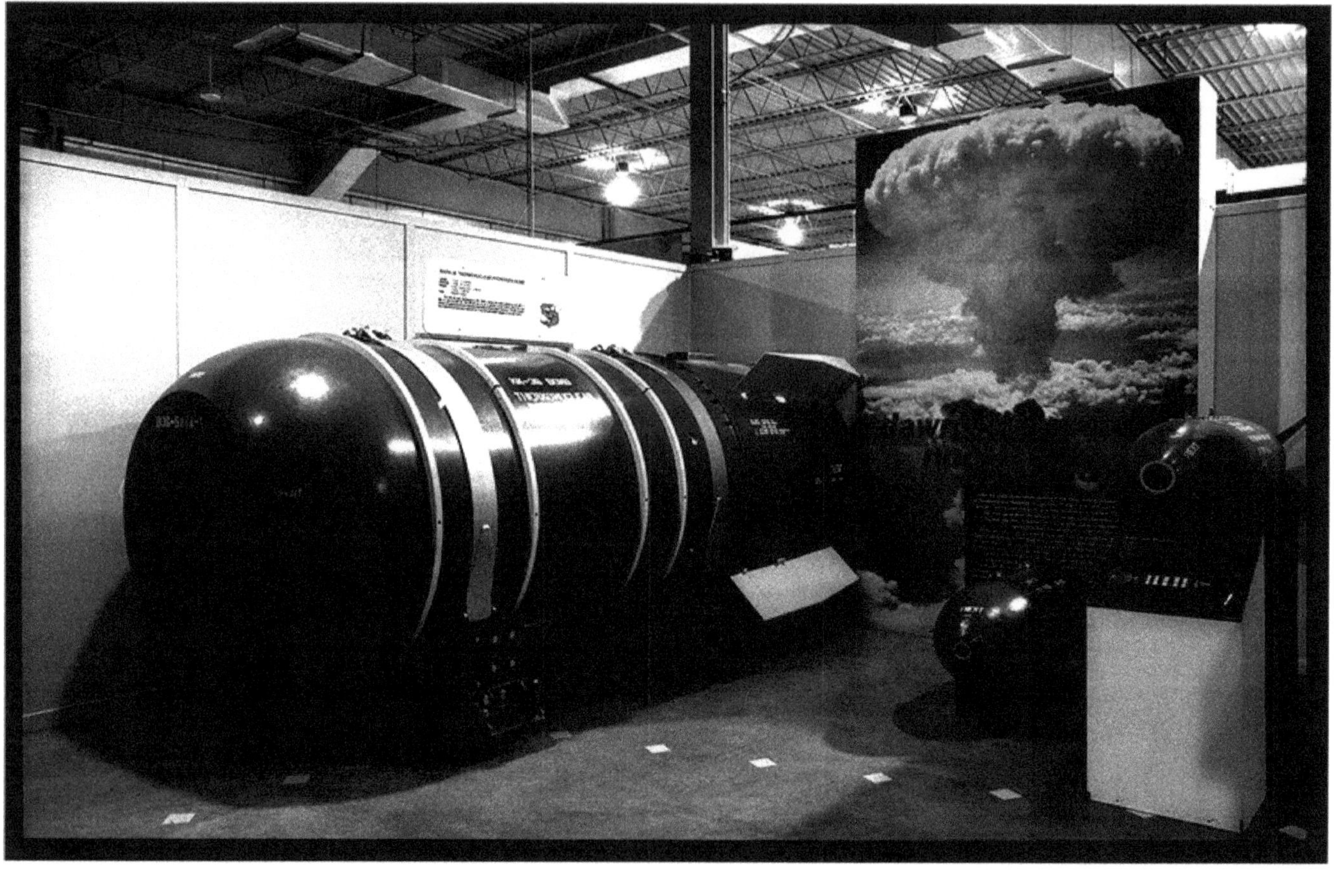

The Mark 36 Hydrogen Bomb was the most powerful weapon in the US arsenal from1956-1961. Over 12 feet long, weighing in at more than eight and a half tons, the Mark 36 had a yield of up to 10 megatons. 920 Mark 36s were made. In 1961 it was replaced by some 500 Mark 41s, the only three-stage fission-fusion-fusion-fission thermonuclear weapon in the US arsenal. Mark 41 was never tested: its projected yield of up to 25 megatons made it the most powerful H-bomb in the US. The Soviets made their own 3-stage H-bomb (the "Tsar Bomba") and tested it at Novaya Zemlya on October 30, 1961 in the largest man-made explosion in history. Its yield was 58 megatons, about 1500 times the combined power of the Hiroshima and Nagasaki bombs. *Offutt AFB, Bellevue, Nebraska. (data: Wikipedia: Mark 36, Mark 41, Tsar Bomba)*

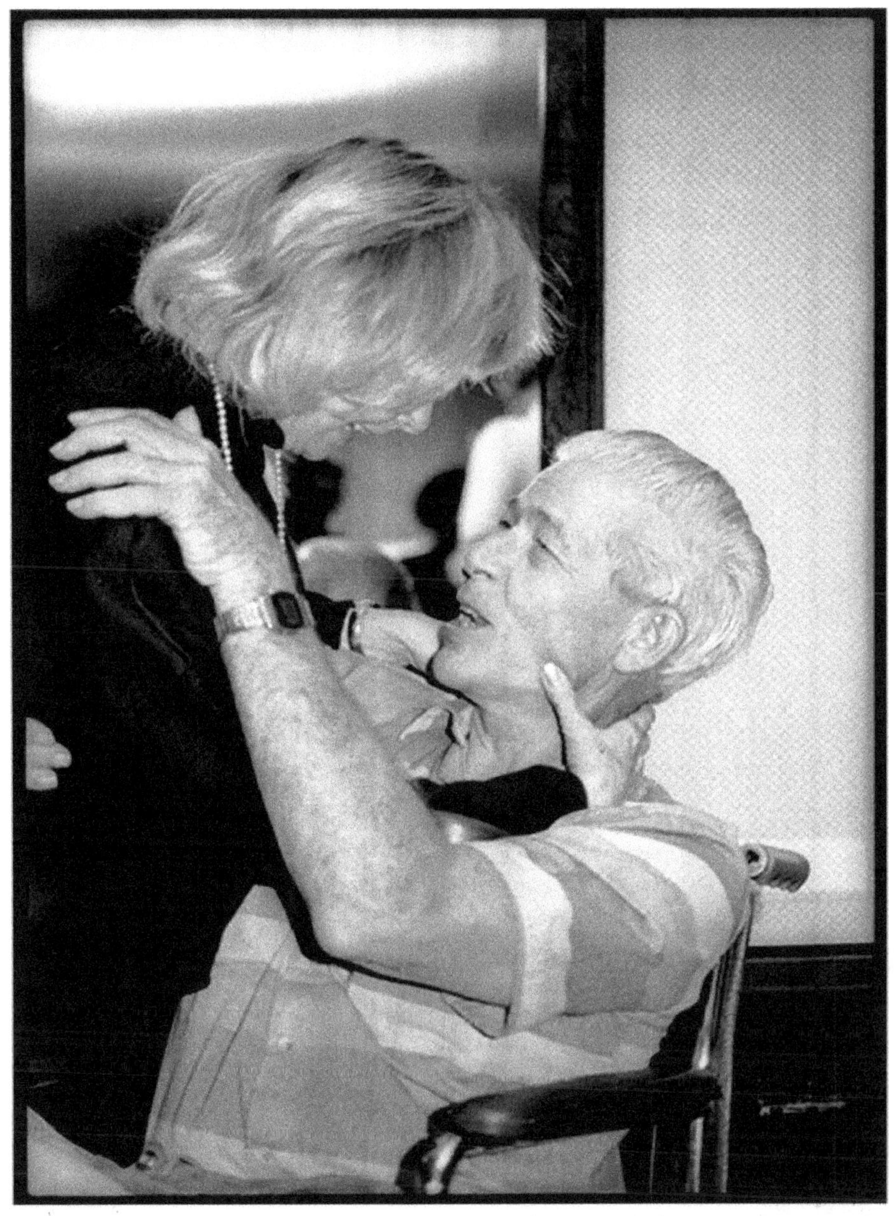

Radiation Activists Pat Broudy and Larry Pray. Pat Broudy is a widow whose husband was an atomic veteran. Larry Pray is an atomic veteran who, in 1952, participated in Desert Rock IV, an exercise involving 7340 soldiers during the Tumbler-Snapper atomic tests in Nevada. That series showcased four air-drops and four tower shots. After the tests Larry Pray experienced loose teeth and bleeding gums. He later developed a brain tumor. *National Association of Radiation Survivors Convention. St Louis, Missouri, 1987.*

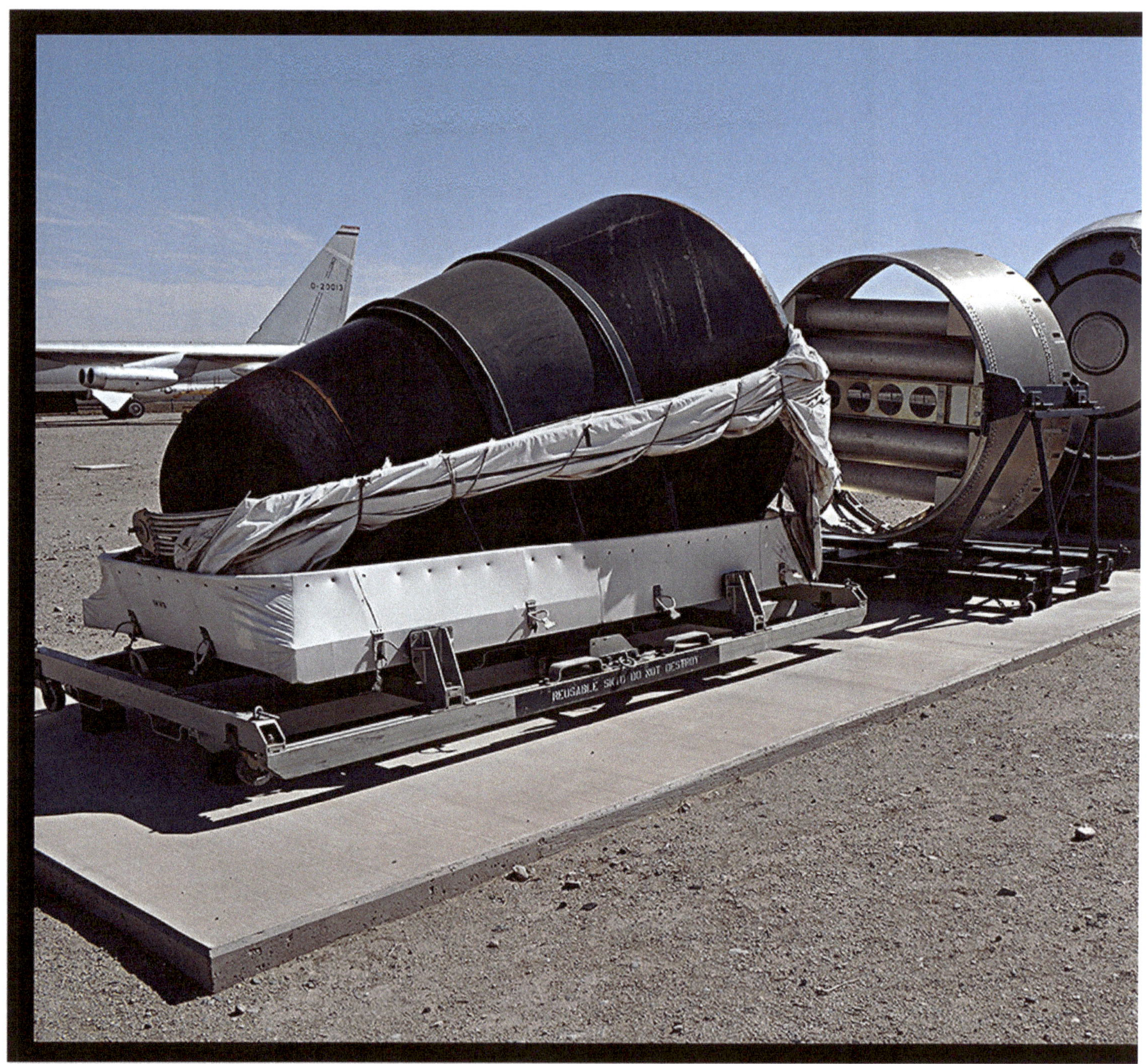

Sliced and Diced: A deactivated Titan, one of the six missiles left out of the original fleet of 108—slowly and History. The Titan II carried the largest warhead of any US delivery system and dominated the arms race

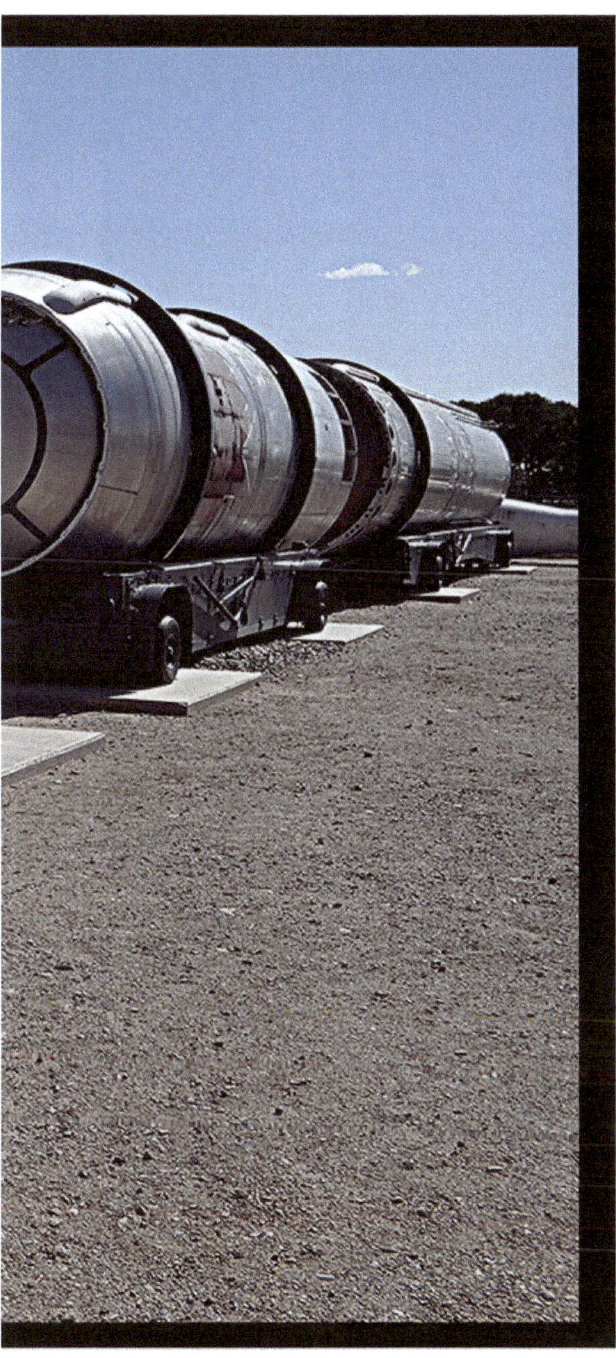

fades from memory on the grounds of a missile park next to Albuquerque's National Museum of Nuclear Science from 1971 to 1986. Aging seals and fuel volatility led to its retirement. *Albuquerque, New Mexico, 2013.*

On Nuclear Seeing

Robert Del Tredici

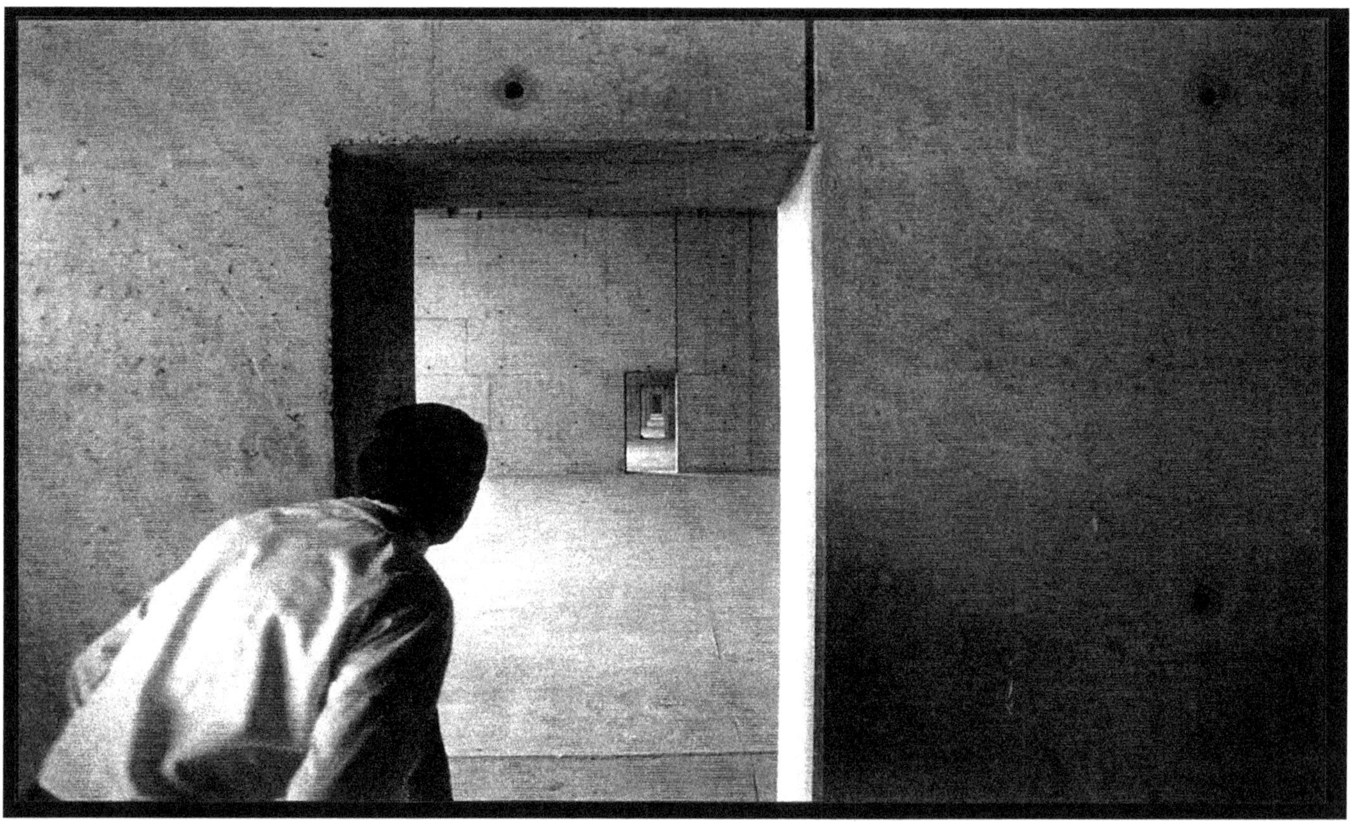

Looking and seeing enable us to make our way through the world. They help us find patterns, spy clues, sight cracks, watch out, and stay sharp. Then there is looking into the nuclear world. This requires a special kind of seeing.

E Area Vault for 300-year storage of low-level radioactive waste, Solid Waste Management Division, Savannah River Site, South Carolina, 7 January 1994.

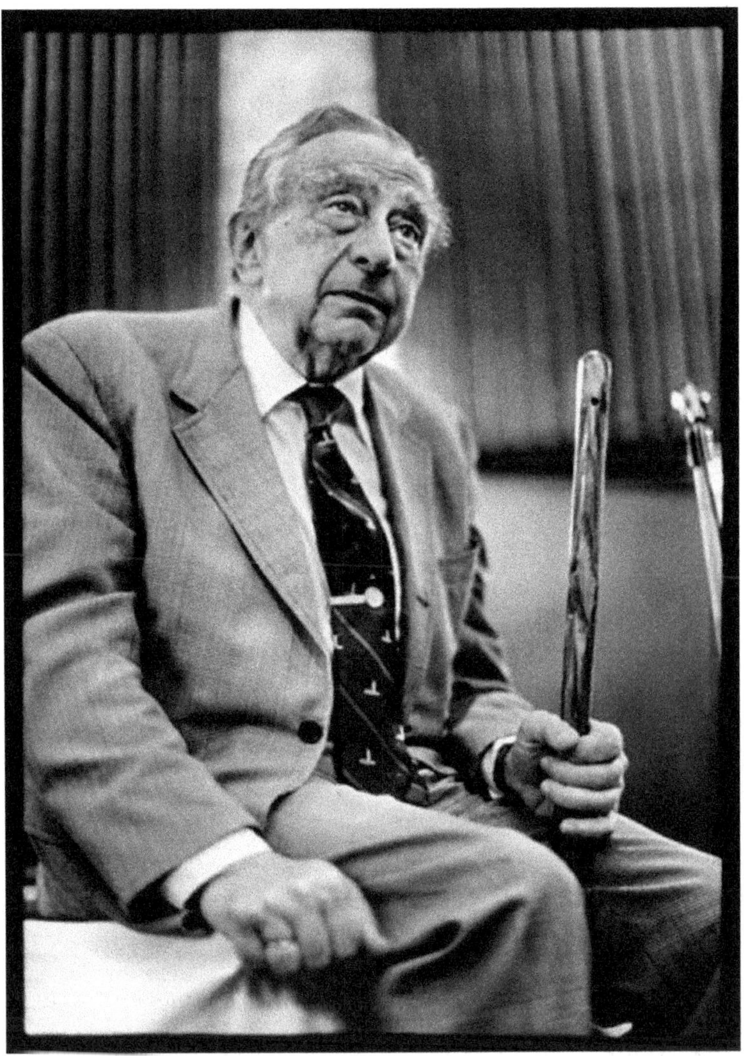

Dr. Edward Teller
"You must never refer to me as the father of **anything**," Edward Teller tells me in Palo Alto on December 21, 1984. A year and a half later, at a public meeting in Montreal, Dr. Teller answers a question about fallout from Chernobyl with the words: "If you explode all the nuclear weapons in all the arsenals of the world, a large fraction of humanity will not be injured, and the damage will be more or less confined to the places where the hostilities occurred." Edward Teller is the Father of the H-bomb. *Montreal, 3 May 1986.*

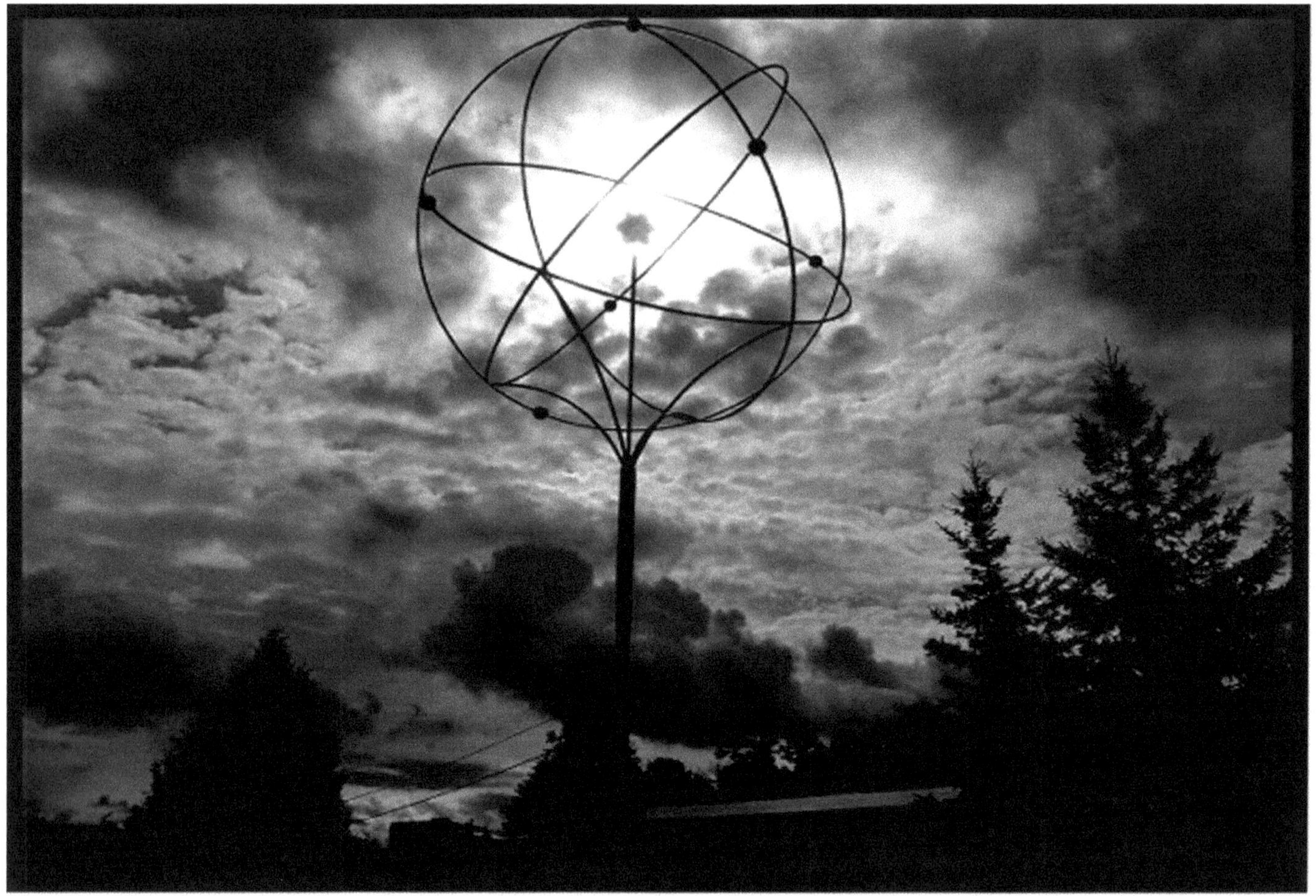

Seeing things nuclear requires an act of the imagination. This structure symbolizes the uranium atom. Uranium is the heaviest naturally occurring element on earth. It is the key element for all nuclear technologies, both civilian and military. And it is a shape-shifter: uranium atoms spontaneously throw off energy and particles, transforming the original uranium into two dozen different elements that are themselves shape-shifters. Uranium has another surprising property: its atoms, alone in all of nature, can be split apart, releasing titanic forces able to be harnessed into the energy of a Bomb. This leads to yet another kind of shape-shifting as these fragments of split atoms are hundreds of other unstable, shape-shifting elements.

Downtown Elliot Lake, Ontario, once the uranium capital of Canada, 16 July 2011.

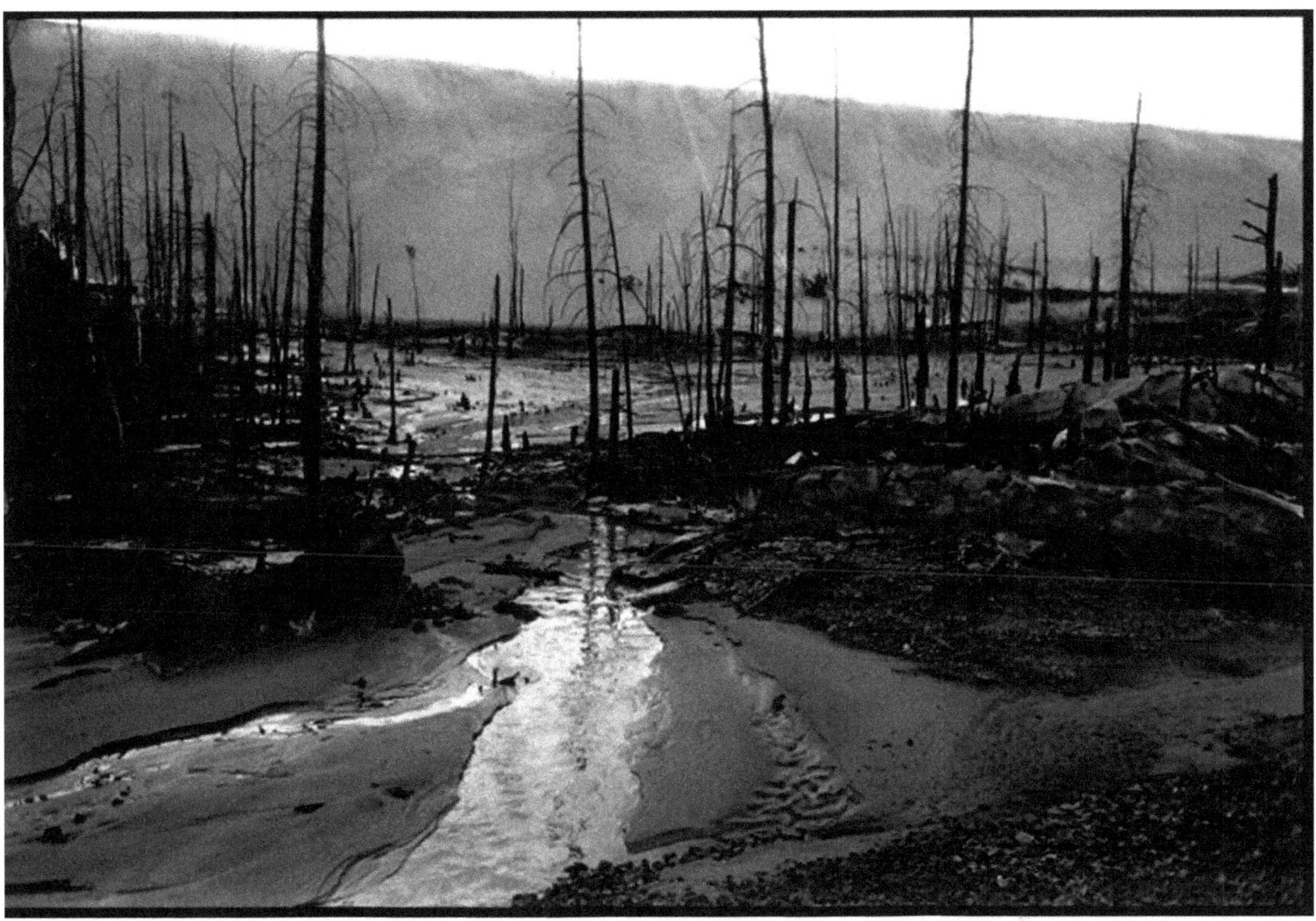

Not far from the town of Elliot Lake stands a wall of radioactive sand thirty feet high made of uranium mill wastes. Behind it is a waterless lake filled to the brim with 70 million tons of these wastes. They contain some of the most deadly materials known to science—radioactive thorium, radium, radon, polonium, bismuth and lead—elements that change from solids to gas and back to solids, whose half-lives range from less than 1/1000 of a second to more than seventy thousand years. All of them are byproducts of uranium that was once locked up in rock underground. The wastes must be isolated from the environment for more than a million years. They were dumped here in the 1950s and early 1960s when Canada was shipping all its uranium south for use in the US Bomb program. The Elliot Lake Chamber of Commerce calls its town "A Jewel in the Wilderness." But the dozens of uranium mines in the region have turned the town's surroundings into an everlasting mess.

Stanrock Tailings Pile, Elliot Lake, Ontario, Canada, 25 August 1986.

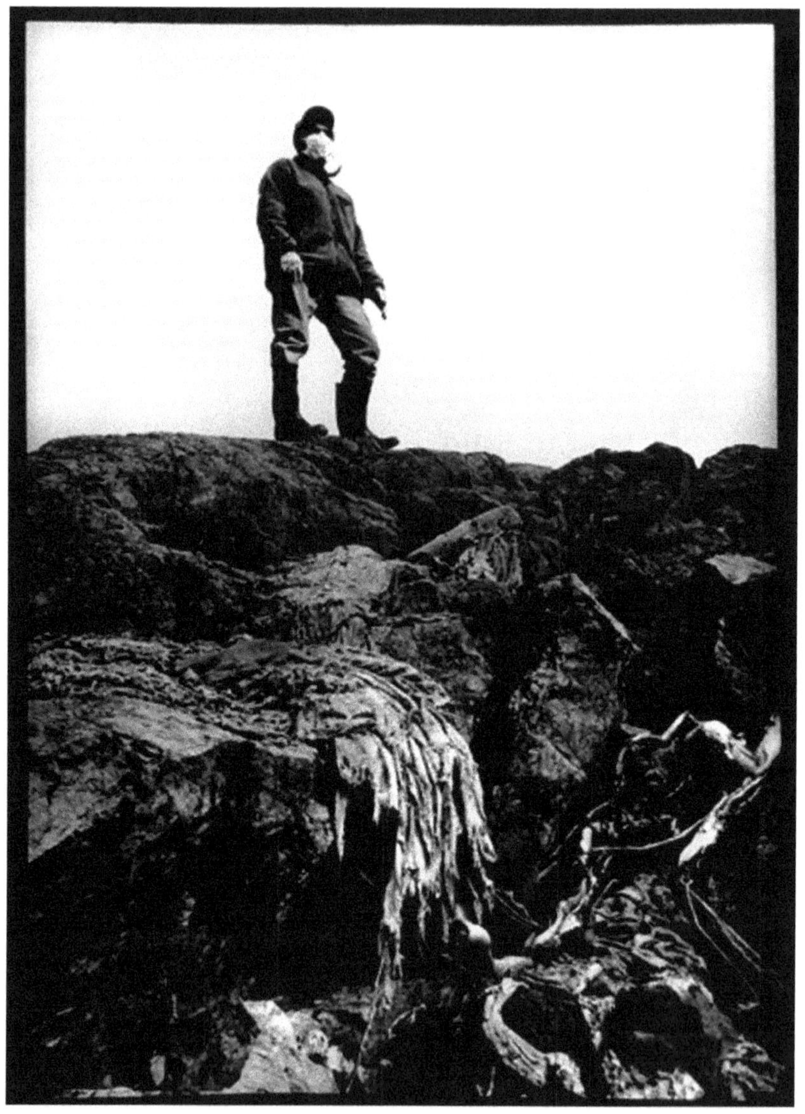

Abandoned Uranium Ore Sacks once filled with crushed radioactive ore and carried by Dene men, now lie rotting in the sun at Port Radium. Dene native Joe Blondin Jr. surveys the scene. He was born in Port Radium when the mine supplied uranium for the first atomic bombs. Many ore-carriers sickened and died; their village on the west shore of Great Bear Lake has come to be known as "the village of widows." *Port Radium, Vance Peninsula, Great Bear Lake, 18 July 1998.*

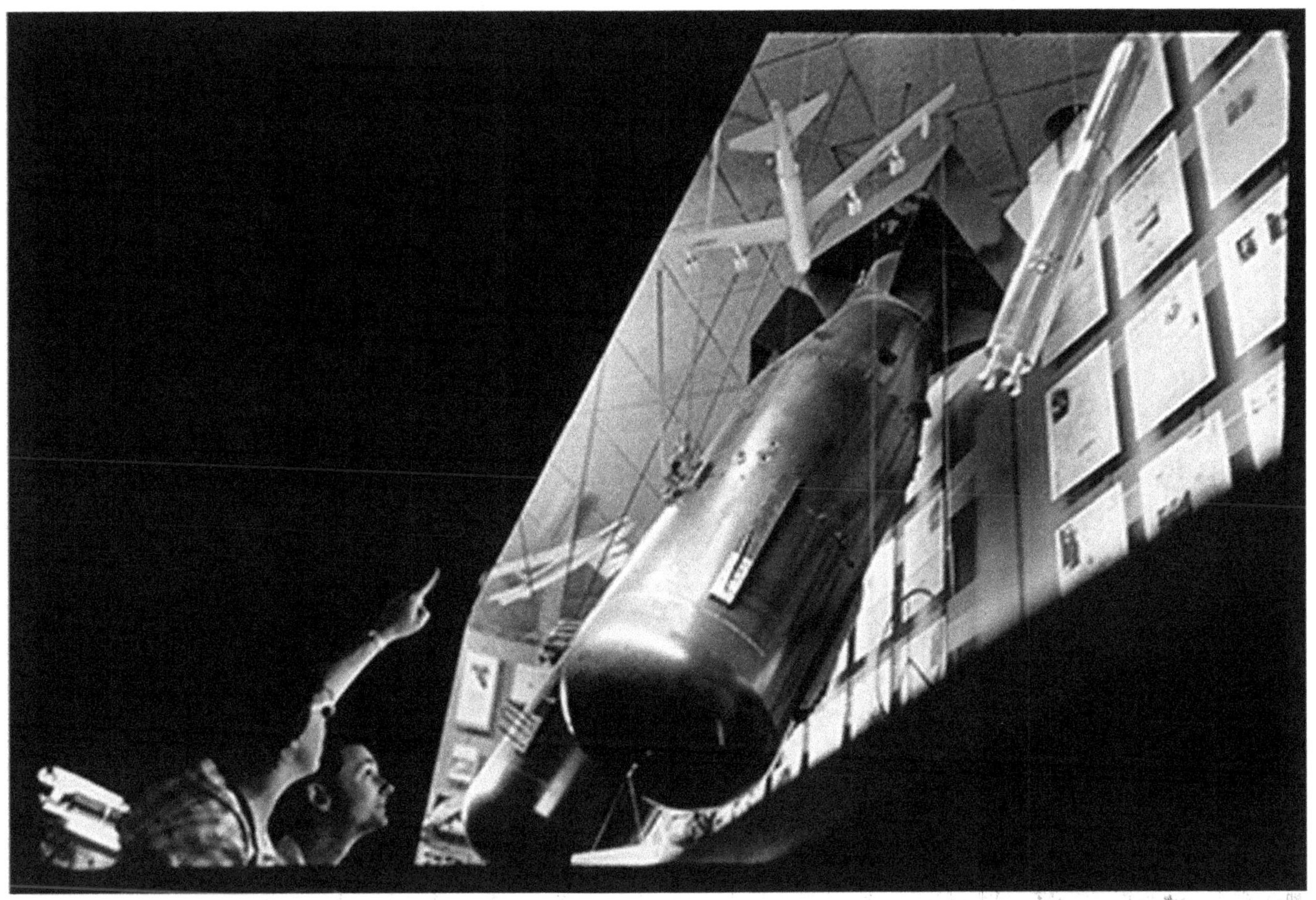

The Hiroshima Bomb. This is a replica of the Hiroshima atomic bomb at the National Air and Space Museum in Washington D.C. It is part of an exhibit entitled The Social Impact of Flight.

In the late 19th century the Dene prophet Louis Ayha had a dream vision about a great bird that carried a great log away from Great Bear Lake and dropped it on the far side of the world where it created fires that could never be put out. *Smithsonian Air and Space Museum, Washington D.C., 25 June 1981.*

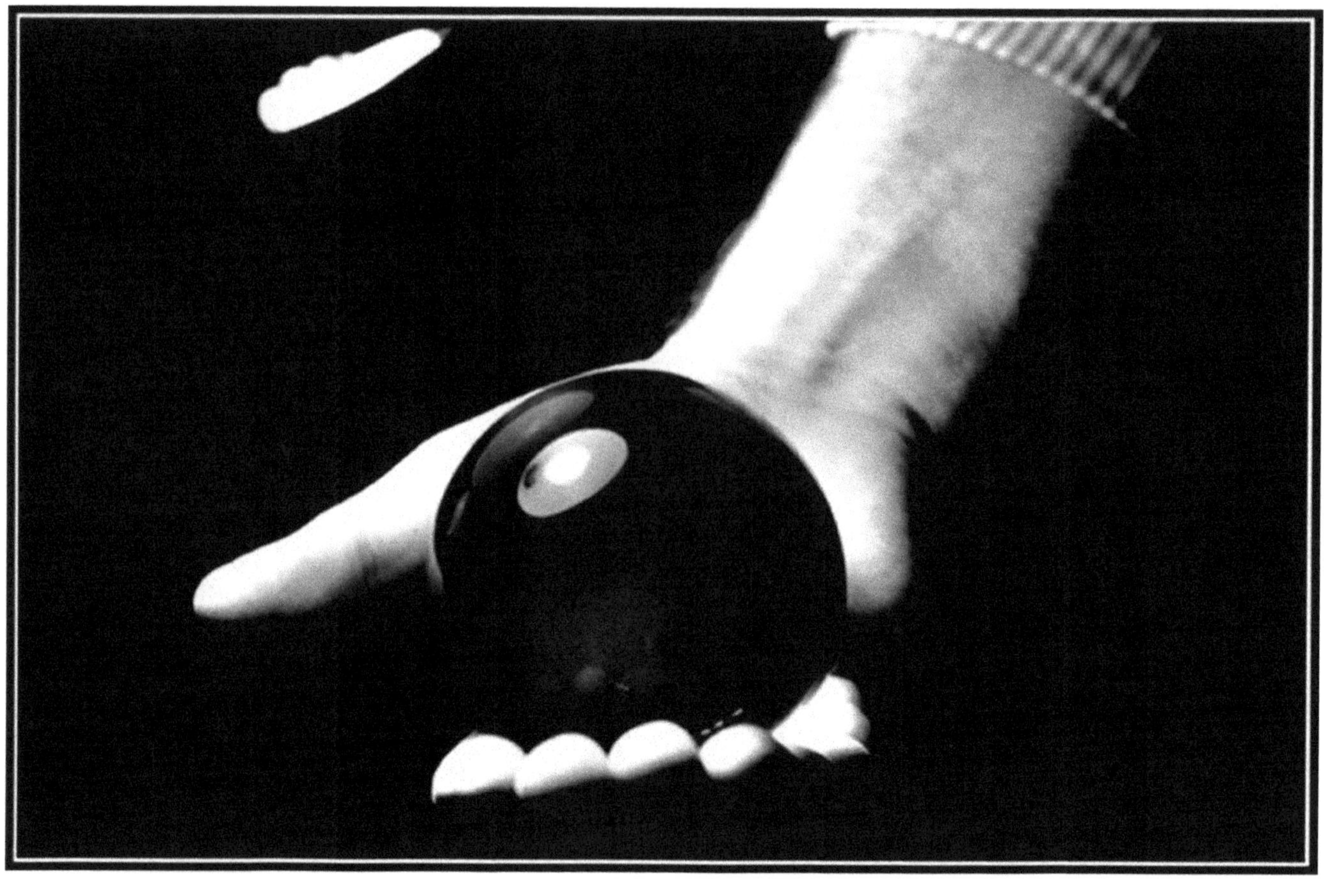

The explosive metal of choice for the Bomb is not uranium but plutonium. Plutonium is automatically created from uranium inside every nuclear reactor. This glass paperweight was designed to match the exact amount of plutonium in the Nagasaki Bomb. Only a fraction of it actually exploded: one penny's worth, by volume, fissioned over Nagasaki to destroy that city.

Glass sphere designed and held by Richard Rhodes, author of The Making of the Atomic Bomb *(Simon & Schuster, 1986). Kansas City, Missouri, 22 September 1983.*

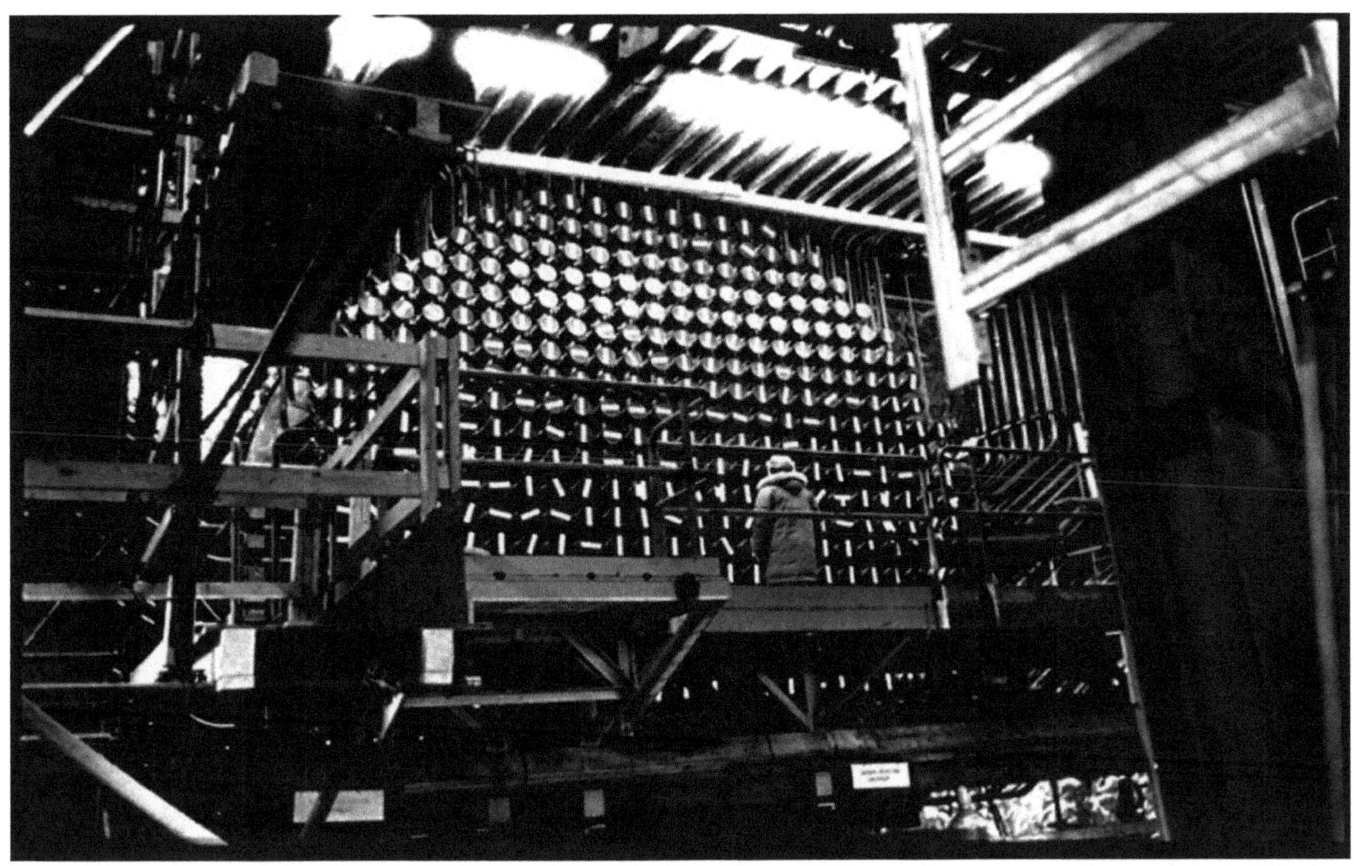

The peaceful CANDU reactor makes plutonium more efficiently than any other power reactor except the breeder. When Canada gave India a CANDU prototype research reactor, India researched its way to a plutonium bomb and its modern nuclear arsenal. India, Pakistan, South Korea, Argentina, Romania, and China have CANDU reactors.

Darlington, Ontario, Canada, 21 January 1987.

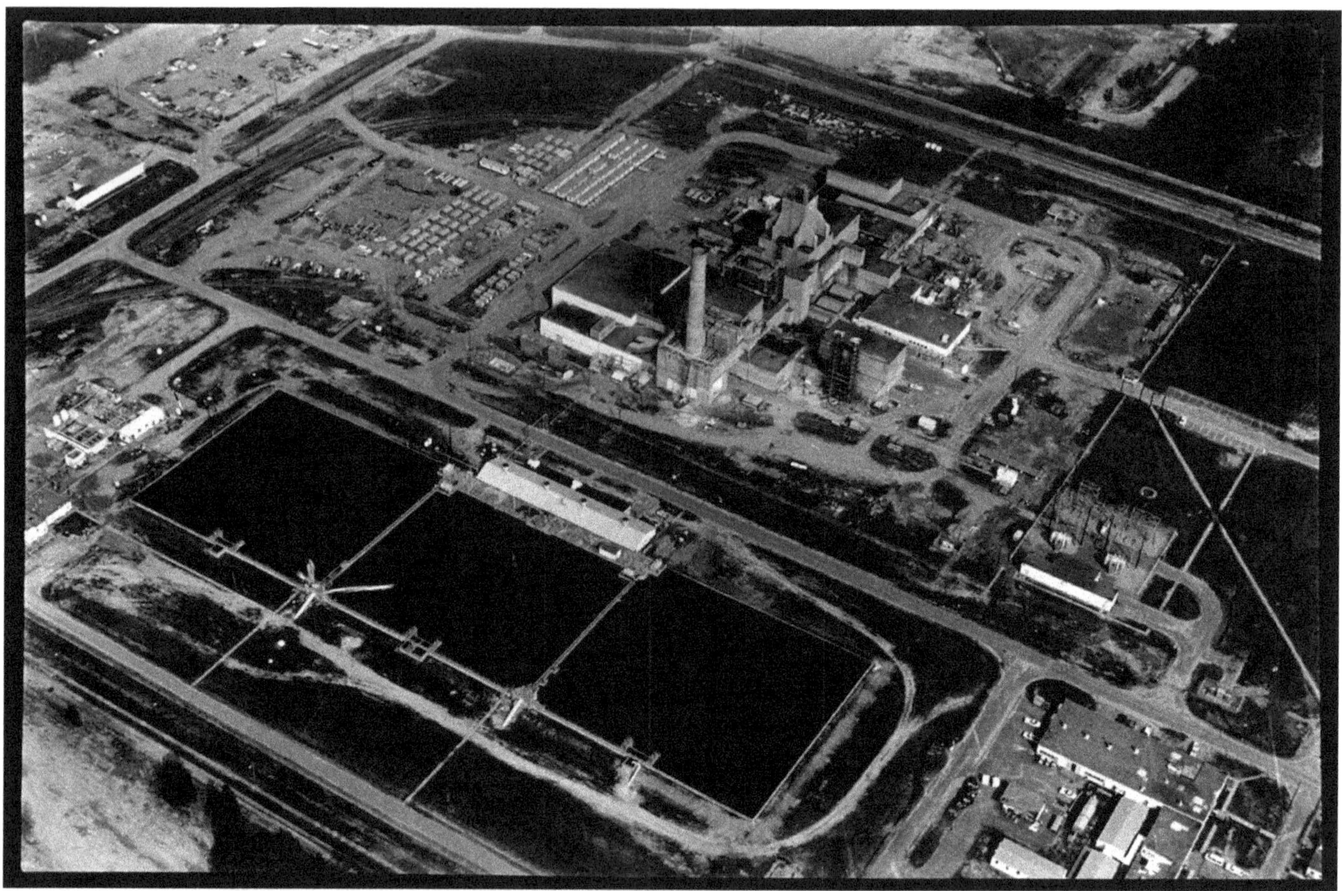

Putonium production reactor at the Savannah River Site. All reactors make plutonium, but this reactor has been specially designed to mass-produce plutonium for bombs. Plutonium is created by neutrons bombarding a uranium metal target; when a uranium atom absorbs a neutron it is transformed into plutonium. The plutonium-rich metal is then chopped up and dissolved in acid, and its plutonium atoms are plated out. Left behind is corrosive high-level radioactive liquid waste that will remain hazardous for millennia. To this day engineers do not yet know how to securely contain these wastes.

The L Reactor, one of five plutonium-production reactors on the 300-square–mile Savannah River Site, Aiken County, South Carolina, 6 August 1983.

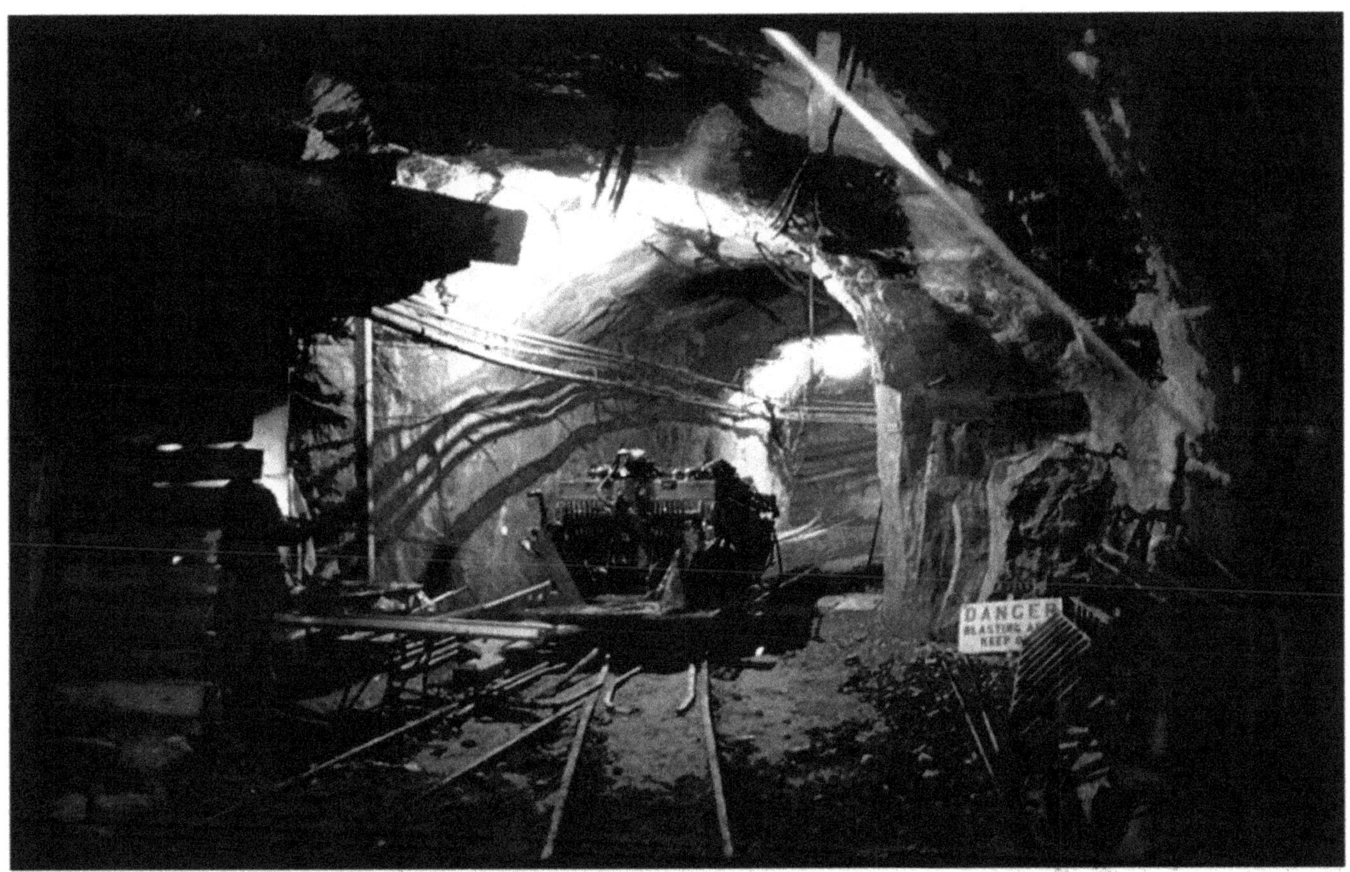

Nuclear advocates describe atomic energy as so clean it can help save the planet. However, they leave out the leftovers from the BEGINNING of the process—the mountains of radioactive uranium mill wastes; they also leave out the product in the MIDDLE—the plutonium created inside all reactors; and they leave out the END product—the high-level nuclear waste that remains deadly for hundreds of millennia, which they still do not know what to do with. The industry's best idea for isolating these wastes has been to bury them deep underground in rock formations that might quite possibly prove to be a perfect solution forever. This test shaft in Manitoba explores the concept of "geologic disposal." The shaft has been sunk a quarter of a mile deep in the stable granite of the Great Canadian Shield, a rock formation undisturbed for billions of years—until this shaft went in.

Tunnel for irretrievable high-level nuclear waste disposal test shaft. Lac du Bonnet, Manitoba, Canada, 15 September 1986.

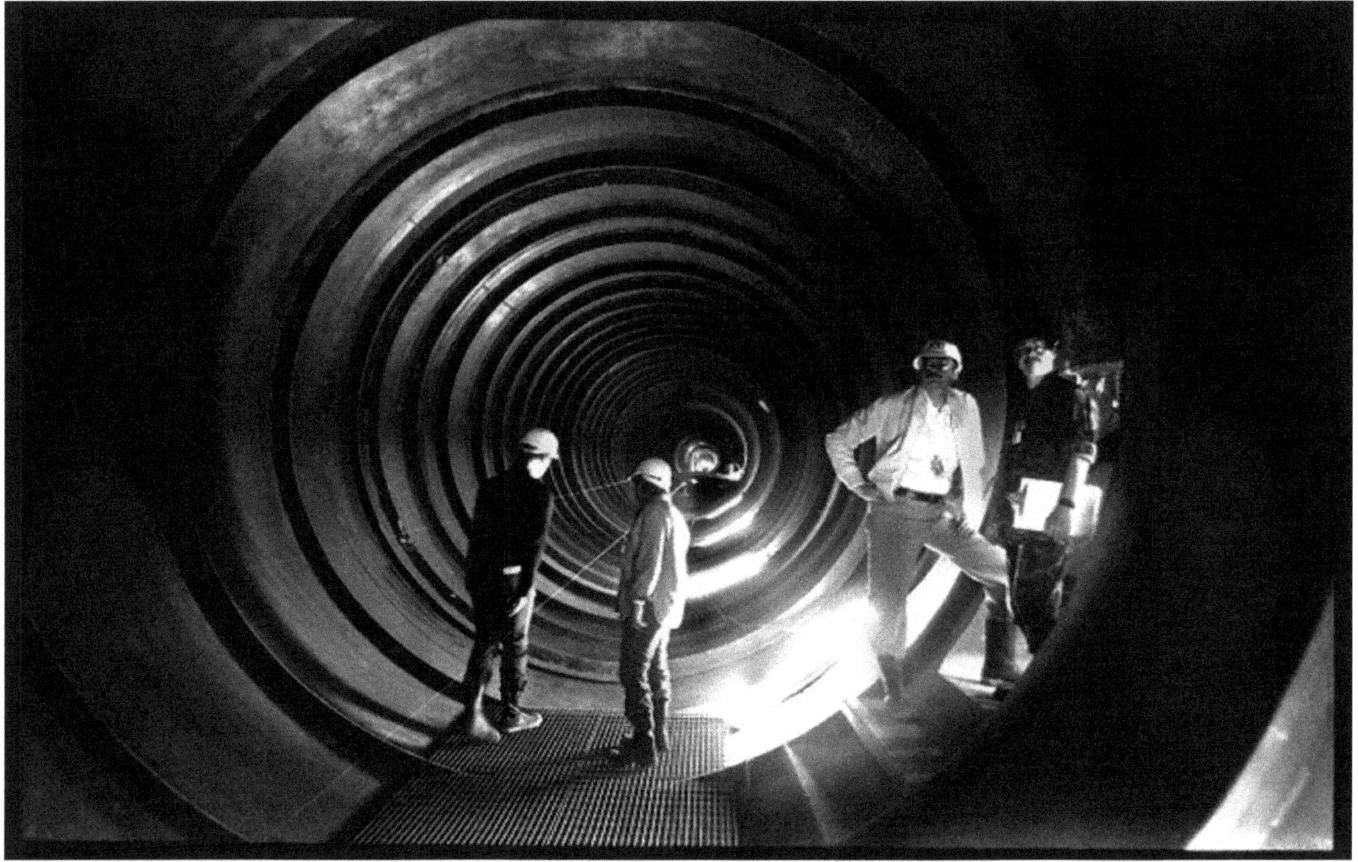

"N" Tunnel. This is an underground tunnel at the Nevada Test Site. It has been designed to capture and analyze the invisible radiation from a nuclear explosion as it impinges on other nuclear warheads at the opposite end of the tunnel. The point is to guarantee that warheads en route to enemy sites, when flooded with radiation from hostile nuclear explosions, will stay the course and arrive at their destinations on target.

It is not nuclear power but the Bomb that drives the nuclear age. Nuclear weapons have burrowed their way deep into our body politic, our planet, and our DNA so that they no longer need an enemy to justify their existence. Our support of their sovereign status fits the profile of an addiction.

Tapered Line-of-Sight Pipe, "N" Tunnel, being readied for nuclear explosion code-named "Misty Rain." Area 12, Nevada Test Site, Nye County, Nevada, 29 October 1984.

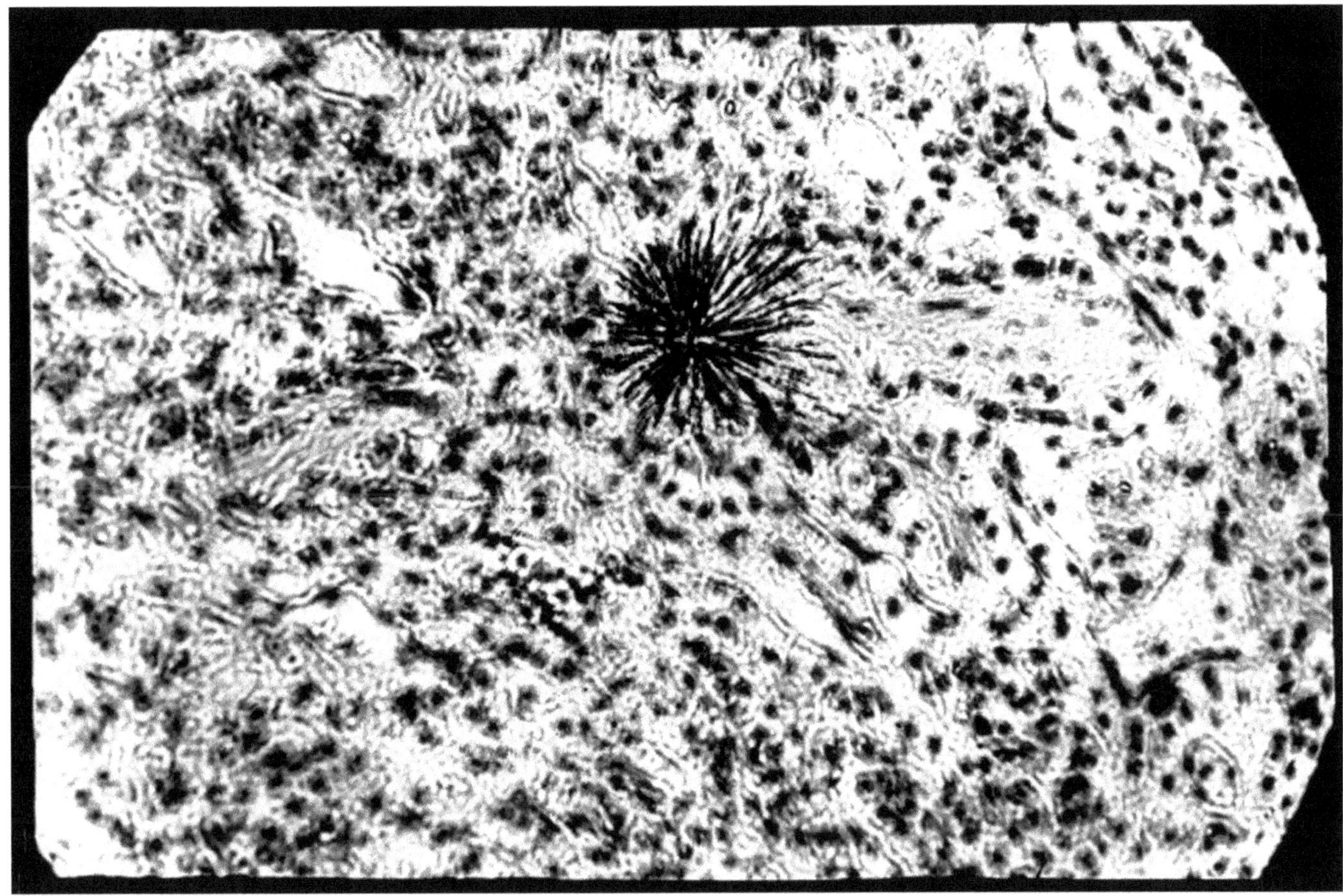

Here we glimpse the radioactive legacy of the Bomb. The black star near the center of the image betrays the presence of a speck of plutonium. We can see the tracks of alpha particles this microscopic speck emits over a 48-hour period. Alpha particles are not very penetrating—they can be stopped by a sheet of paper; but inside the lung, the 10,000 living cells within the range of that black star receive a constant heavy dose. Plutonium has a half-life of 24,400 years.

Slide provided by the Lawrence Radiation Laboratory, Berkeley, California, 20 September 1982.

On the Atomic Photographers Guild

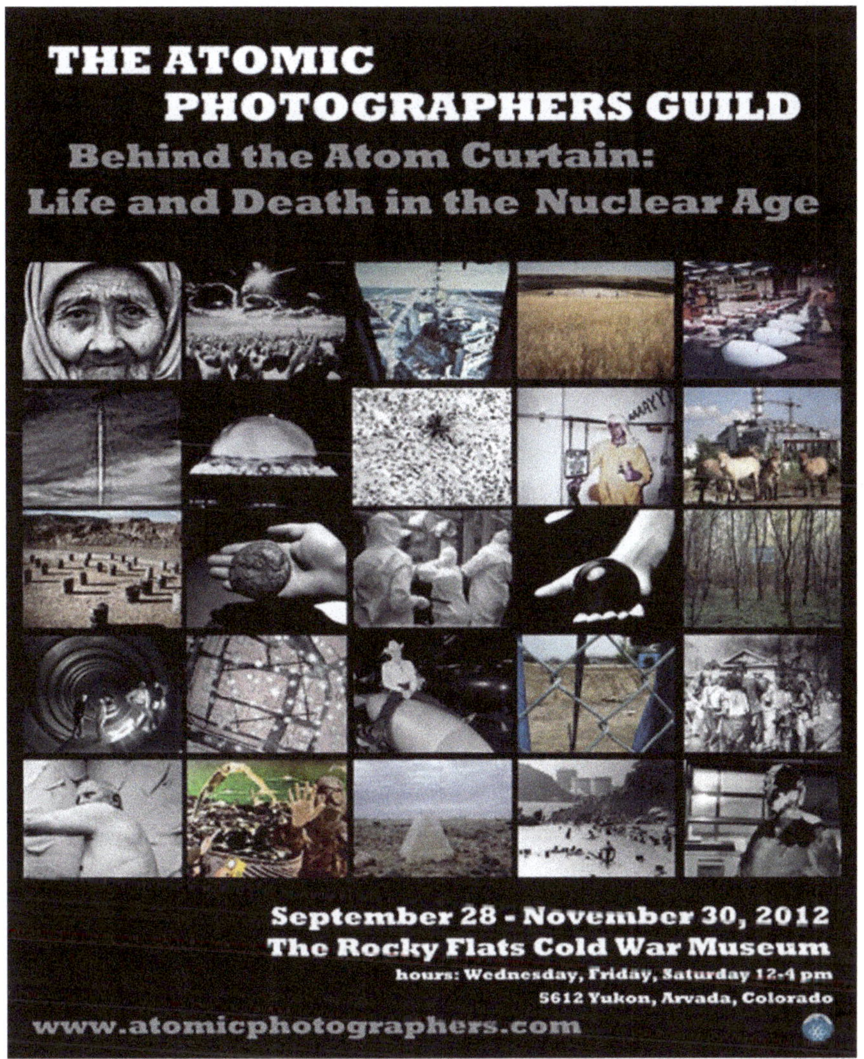

While photographing the Bomb I met other photographers also documenting the nuclear age. In 1987 I founded the Atomic Photographers Guild to bring together colleagues like James Crnkovich from the US, image-makers from Japan, Russia, Germany, South America, Kazakhstan, Australia, and Canada. The Guild currently has 25 members.

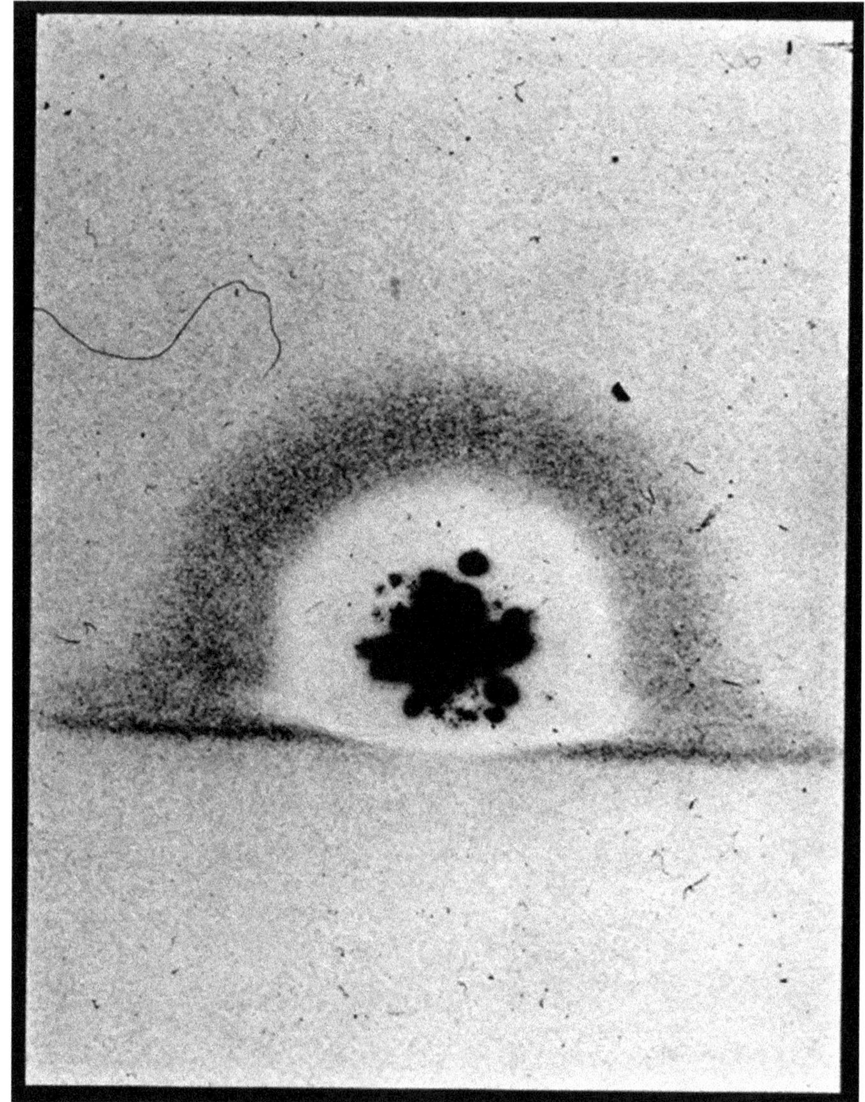

The Guild includes the world's first atomic photographer, Berlyn Brixner, the official photographer of the "Trinity" explosion three weeks before Hiroshima. He deployed 50 still cameras and movie cameras from half a mile to six miles from Ground Zero. This image is from a high-speed motion picture camera that captured the first light off the bomb—a light so brilliant it melted a hole in the film. Thus the instant the Bomb appeared on earth it could not be seen by human eyes. It has remained culturally invisible ever since.

Ground Zero, Trinity blast, Alamogordo Desert, July 16 1945. High-speed motion picture frame capture by Berlyn Brixner. From At Work in the Fields of the Bomb, *Robert Del Tredici (Harper & Row, NY, 1987)*

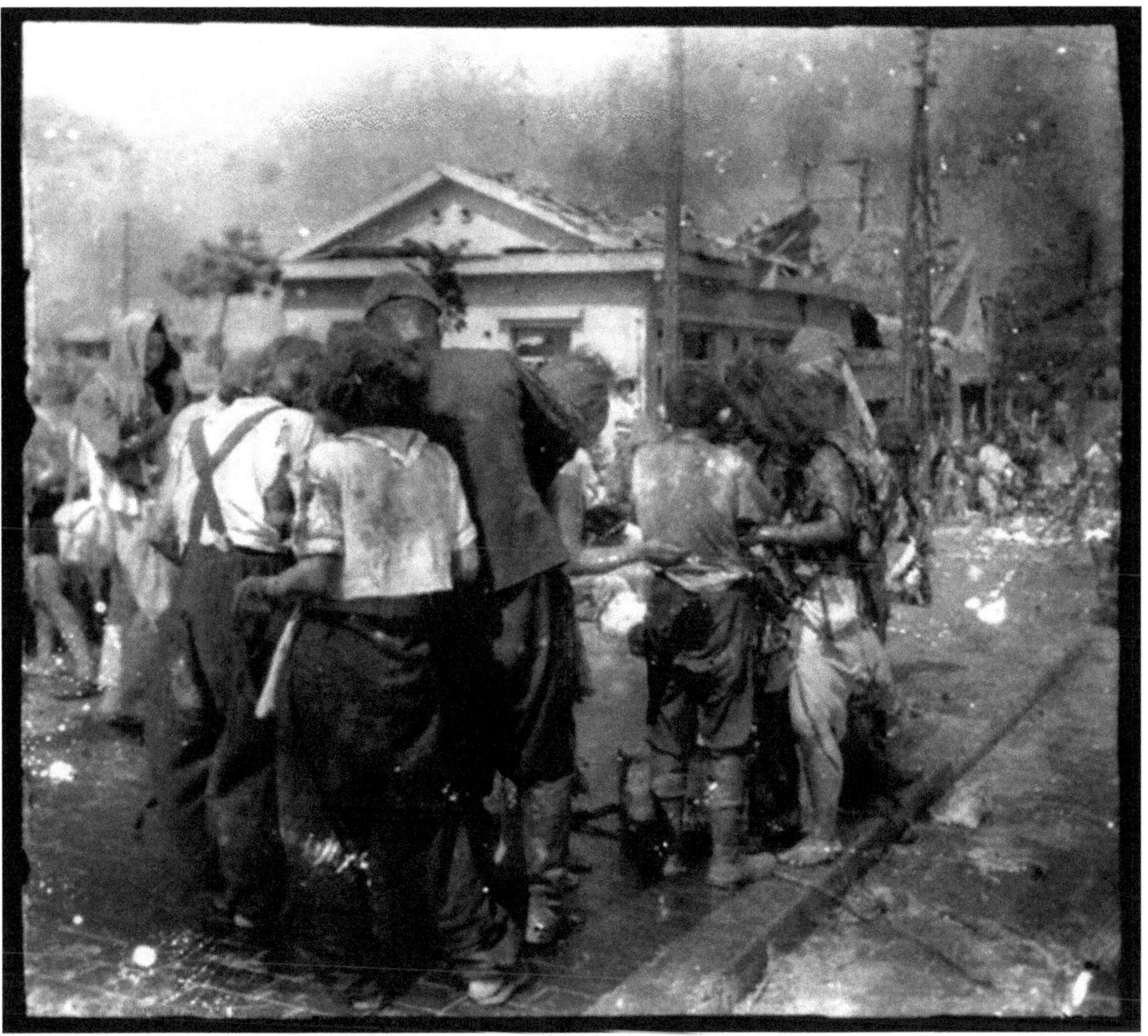

The world's second nuclear photographer is Guild member Yoshito Matsushige, the only photographer to have taken pictures inside Hiroshima the day the A-bomb fell. He had two rolls of film but could not bear to press the shutter more than six times. He came up with five photographs. The pictures are remarkable for what they do not show: we see no corpses or the wounded and dying; we are shown a group of stunned children, people waiting for food, and debris around Matsushige's home. Though impelled by a sense of duty to capture the atomic disaster, on that day Yoshito Matsushige could shoot no more.

Students from the Hiroshima Girls' Commercial High School and the Hiroshima Prefecture Daichi Middle School at the Miuki-bashi Bridge in Hiroshima. Taken by Chiboku-Shimbun photographer Matsushige with 6x6 Mamiya camera. Hiroshima, August 6, 1945. From At Work in the Fields of the Bomb, *Robert Del Tredici (Harper & Row, NY, 1987)*

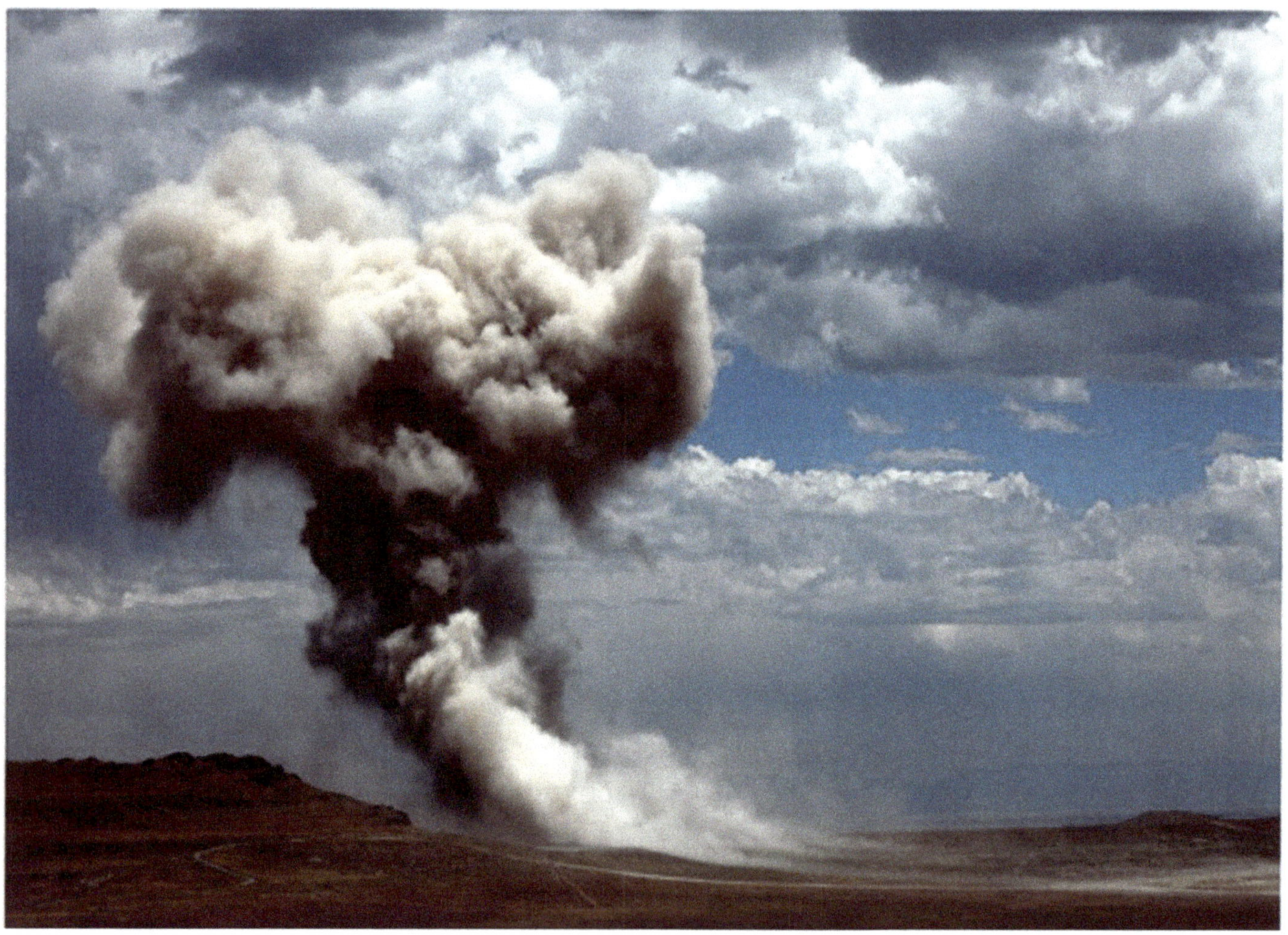

This 3,000 foot tall mushroom cloud was captured by Canadian Atomic Photographer Mary Kavanagh. It is not a nuclear cloud, but it is part of the nuclear story because it is the result of a chemical explosion detonated in accordance with the START treaty to destroy a 20-ton Minuteman Missile Rocket Motor at a U.S. nuclear disarmament site in the Utah desert C-4 detonation of Minuteman Missile Stage One Rocket Motor weighing 41,200 pounds.

Minuteman Intercontinental Ballistic Missile Motor and Ordnance Disposal Area, United States Army, Utah Test and Training Range, North Range, Utah, 2010.

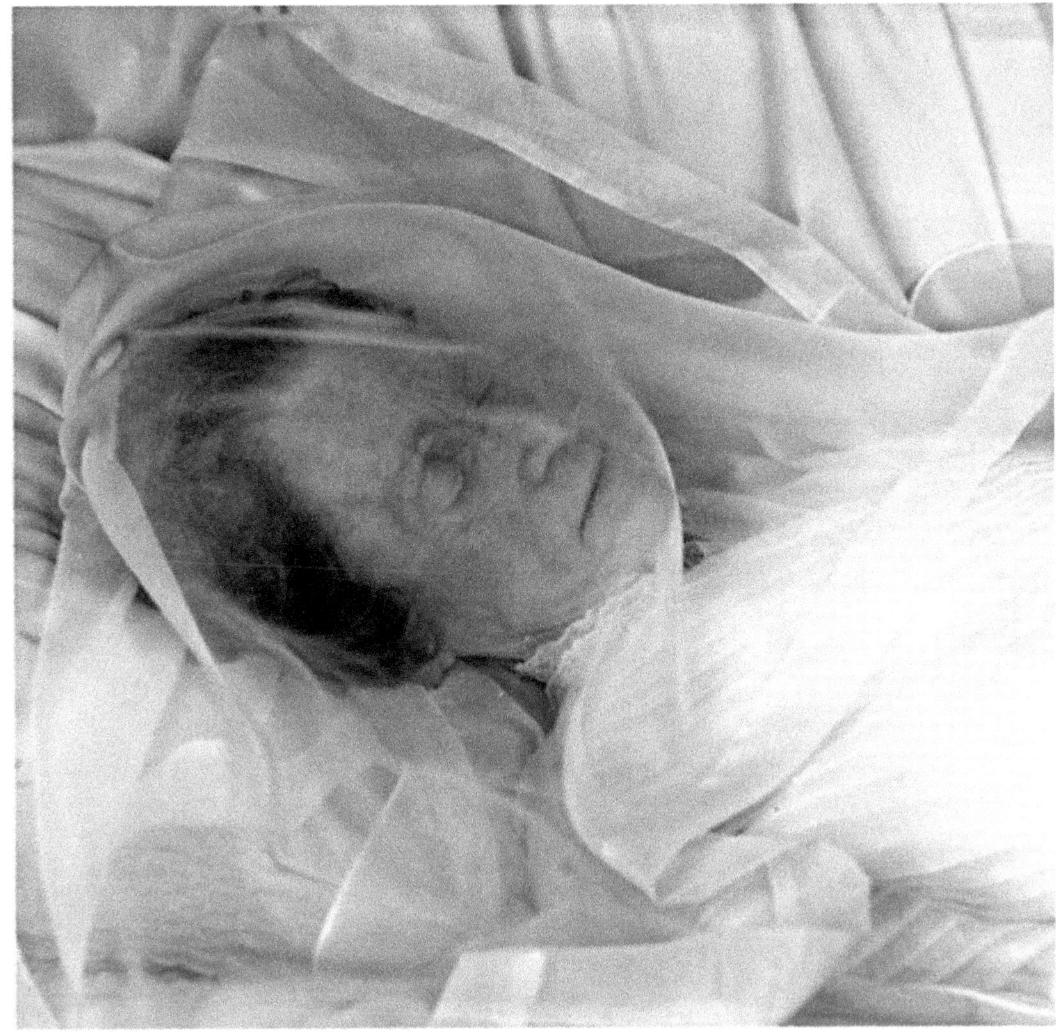

Carole Gallagher (USA) has photographed victims of American nuclear weapons living downwind from the Nevada Test Site in the 1950s and 60s when fission products from atmospheric explosions fell into their lives. The explosions were called "tests" but their impact on surrounding humans was dire. Della Truman, here, is being laid to rest. She had just died of a heart attack from "thyroid storms" in a thyroid that had been damaged by years of exposure to radioactive fallout.

West Jordan, Utah, 1987. Photograph © Carole Gallagher from American Ground Zero: The Secret Nuclear War *(MIT Press, 1994).*

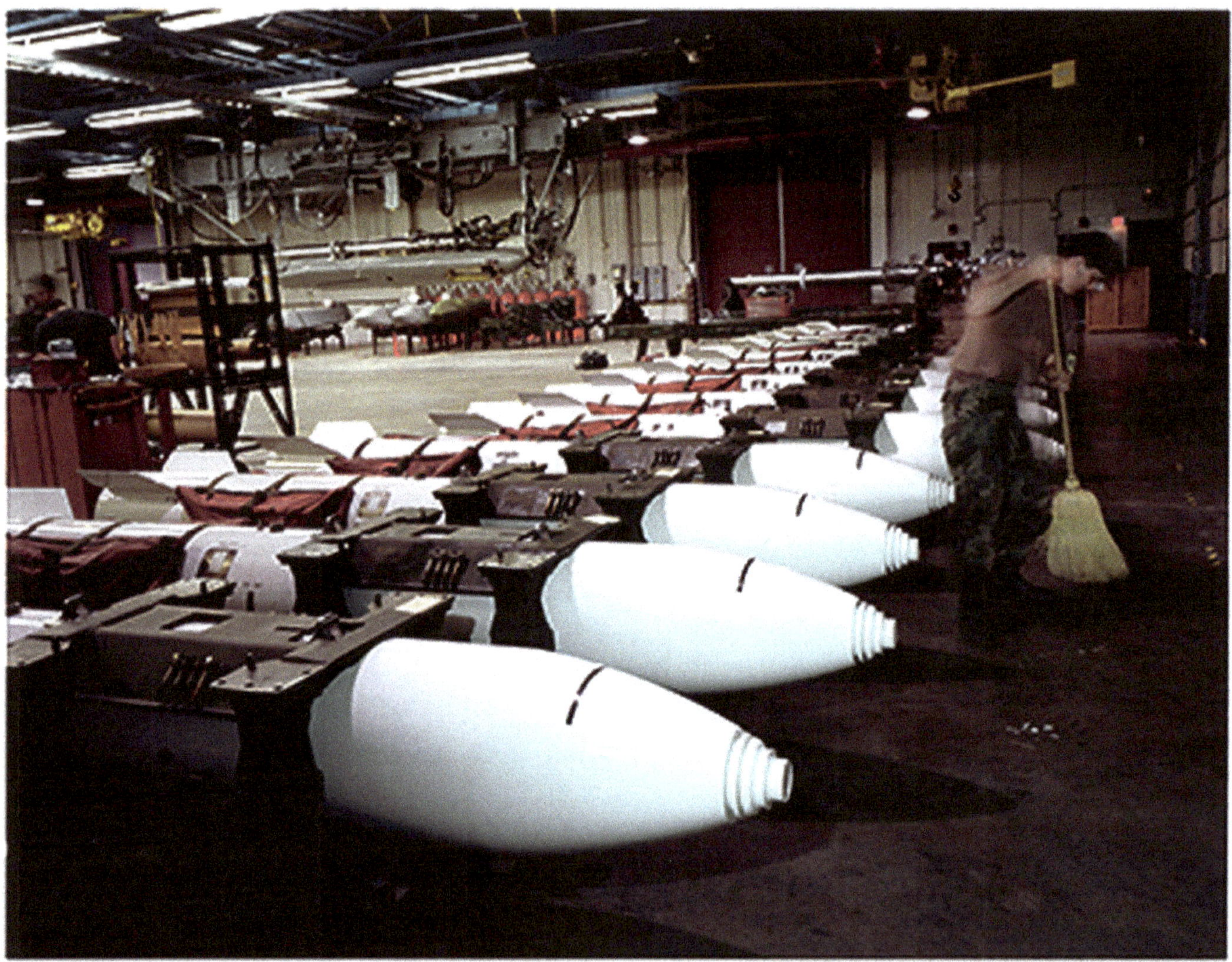

Paul Shambroom (USA) has photographed the deployments of the modern U.S. nuclear arsenal. Here a soldier tidies up a storage room at the Barksdale Air Force Base in Louisiana. This room contains B83 1-megaton nuclear gravity bombs, each with the power of 50 Hiroshima atomic bombs.

Weapons Storage Area, Barksdale Air force Base, Louisiana, 1995. Photograph © by Paul Shambroom, from Face to Face with the Bomb *(Johns Hopkins University Press, 2003)*

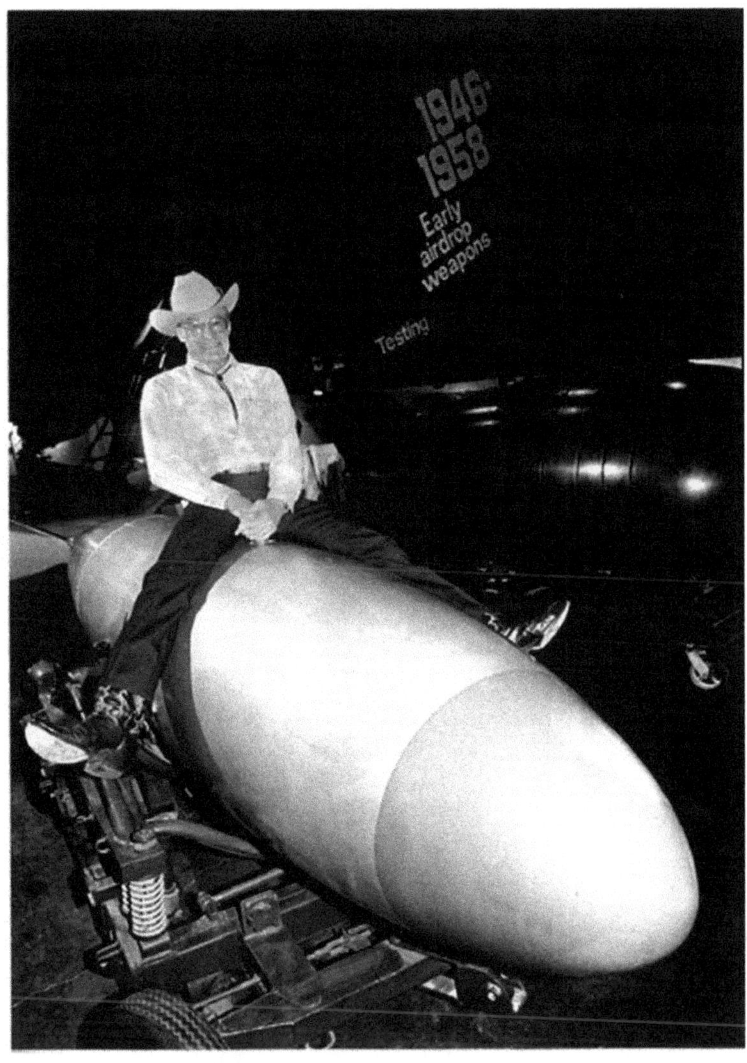

James Lerager of California captured Major Hank Henry atop a 500-kiloton "Super Oralloy" atomic bomb—the largest fission bomb ever dropped by plane (in Operation Ivy, as shot King, an airburst over Einewetok atoll in 1952). Hank was a fighter-bomber pilot, one among the hundreds assigned to the top-secret 4925th Test Group (Atomic), also known as "The Megaton Blasters." These pilots flight-tested Air Force nuclear weapons delivery systems, dropping live nuclear bombs in the Pacific and over the Nevada Test Site—and sometimes flying through the mushroom clouds. The 4925th was operational from 1948 to 1961. At least 73 of its members have died from radiation-related cancers.

Photo © James Lerager. National Atomic Museum reunion of the 4925th Test Group (Atomic), Kirtland Air Force Base, Albuquerque, New Mexico, 1986. From Tales from the Nuclear Age *by James Lerager (1994, City Gallery of Contemporary Art, Raleigh, NC). Stanley Kubrick began work on* Dr. Strangelove *in 1960.*

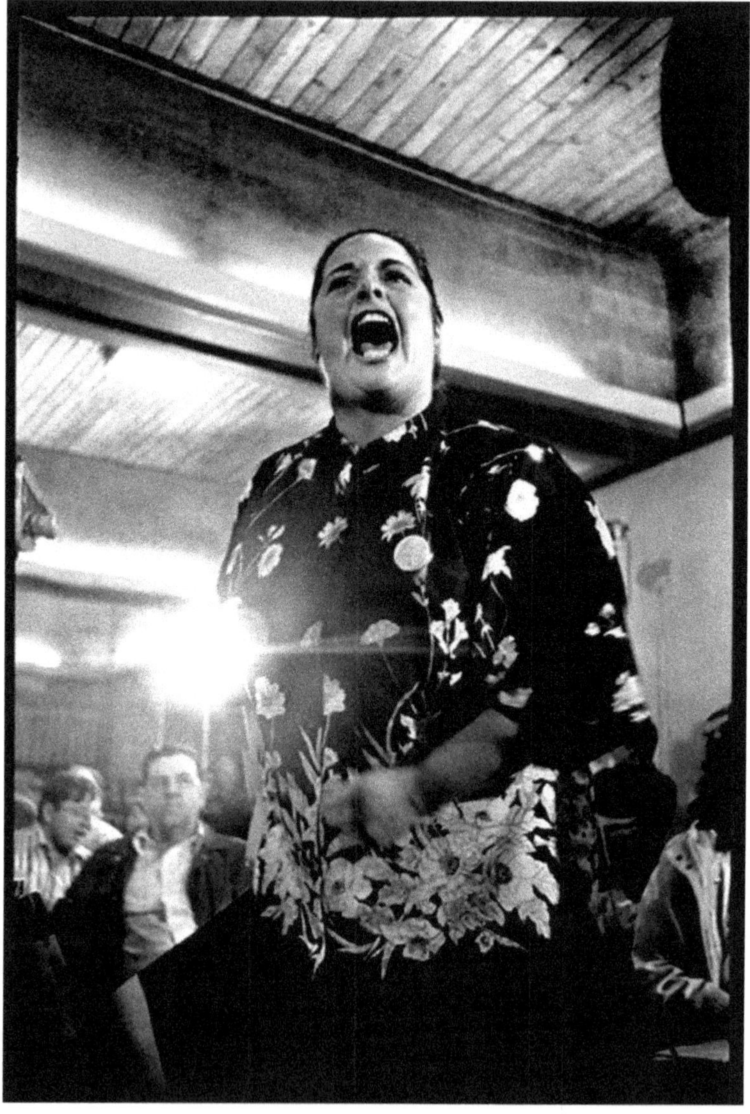

One year after the 1979 partial core meltdown at Three Mile Island, local housewife Joyce Corradi voices opposition to the industry's plan to decontaminate air inside the crippled reactor by releasing radioactive gas into the atmosphere around her town. A costlier option to freeze the gas and ship it off-site was not chosen. In spite of near-total public opposition, the venting went forward. One year later a court of law declared the radioactive release illegal. The partial meltdown at Three Mile Island has been the gravest nuclear accident in U.S. history.

Liberty Fire Hall, Middletown, Pennsylvania. 19 March 1980. Photograph © Robert Del Tredici, from The People of Three Mile Island *(Sierra Club Books, San Francisco, 1980).*

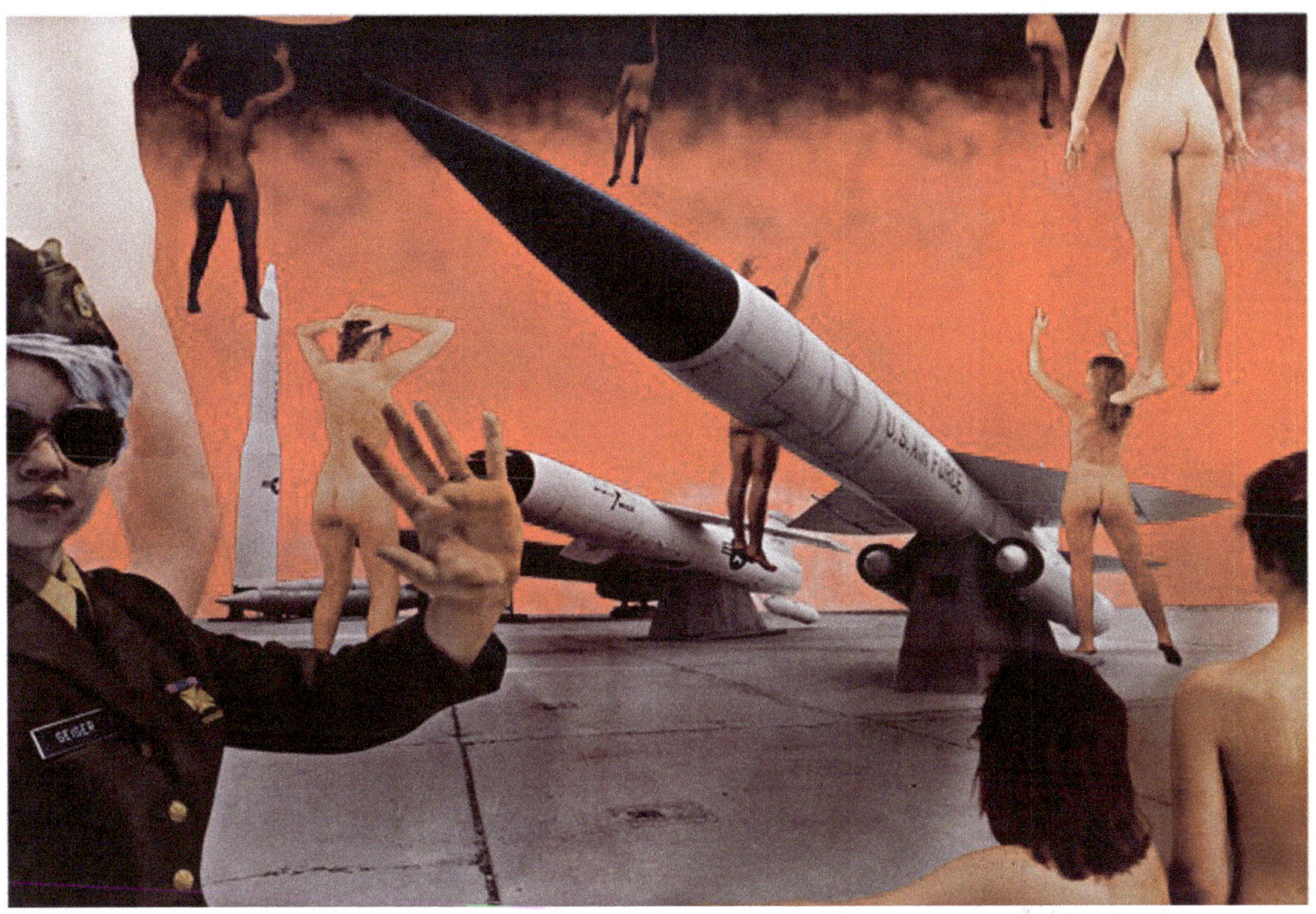

Patrick Nagatani (USA) approaches the Bomb with a mythic eye, using camera and collage. He calls this tableau "Lysistrata" after the 411 B.C. comedy by Aristophanes about women who discontinue sexual union with their men until the men stop going to war. Patrick also calls the image "Women Leaving the Earth."

Photograph © Patrick Ryochi Nagatani from Nuclear Enchantment: Photographs *by Patrick Nagatani, essay by Eugenia Perry (University of New Mexico Press, 1991).*

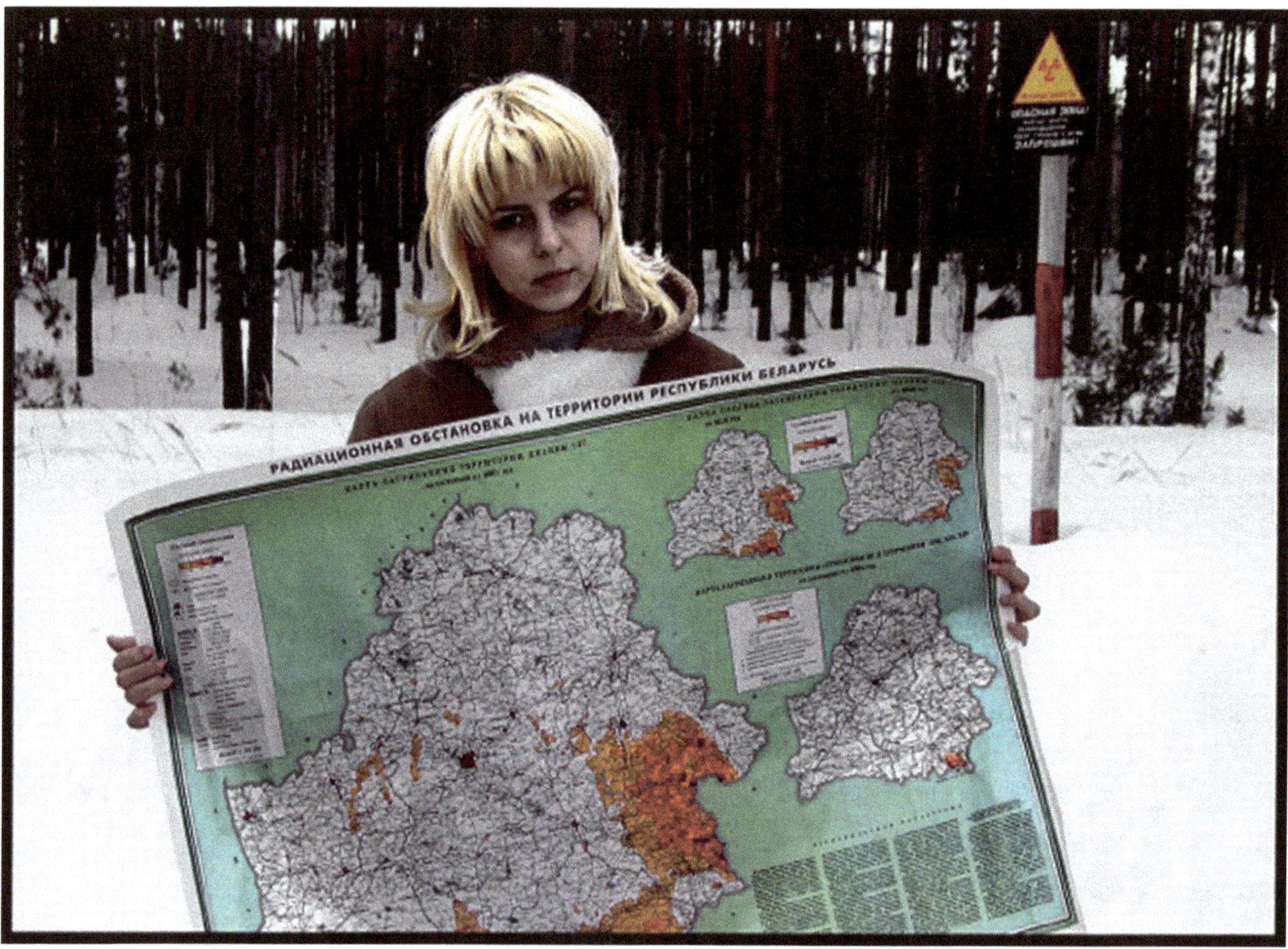

Vaclav Vasku of the Czech Republic has visited the exclusion zone around the Chernobyl reactor many times to document the ongoing aftermath of what was the world's most serious nuclear accident until Fukushima. Eighteen-year-old Julia holds a map showing how over 25% of Belarus has been contaminated by fallout from Chernobyl.

Village of Bykhov, Mogilev, Belarus, 23 March 2005.

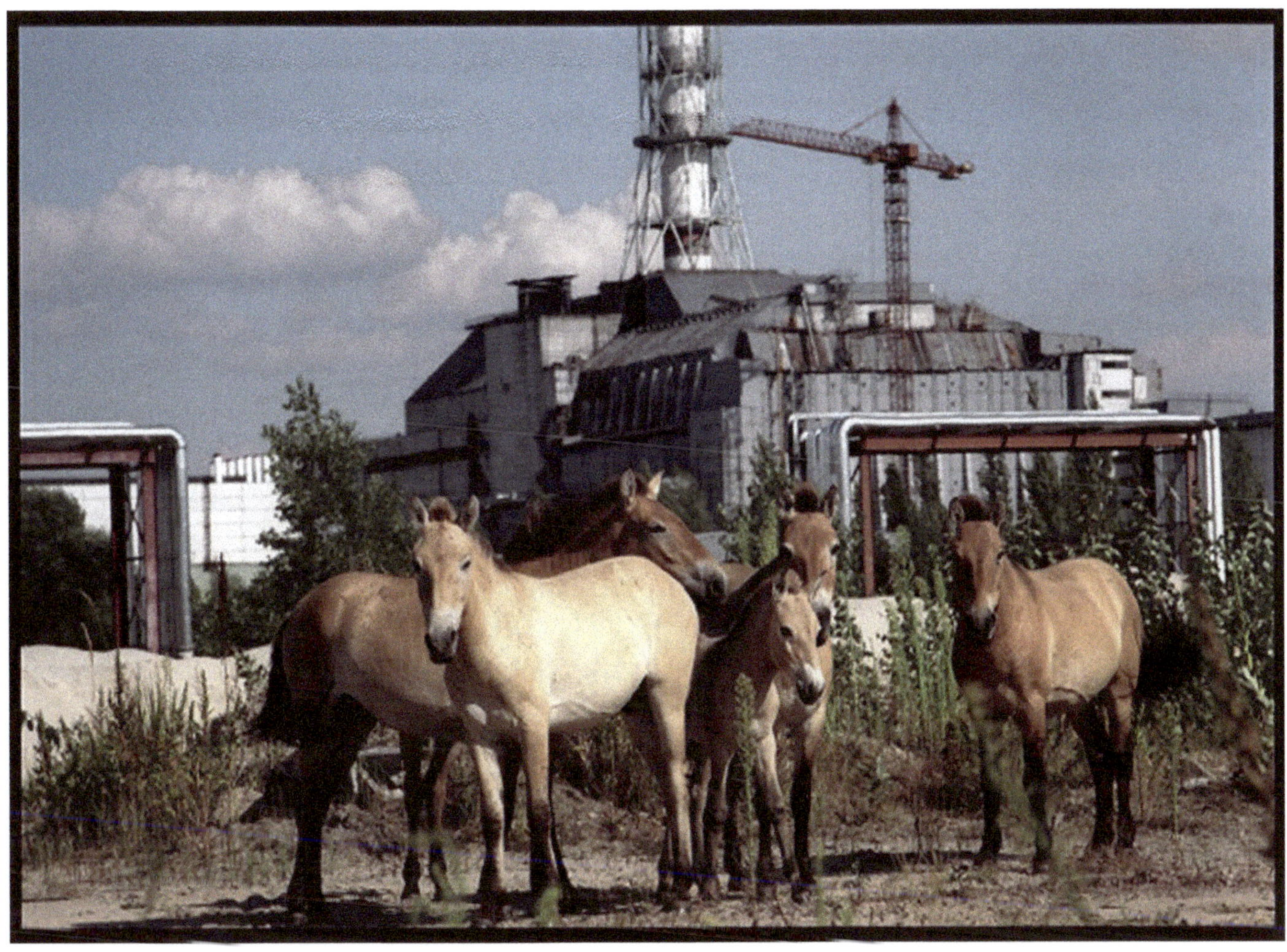

Przevalsky wild horses are the world's oldest horses; we see them on the walls of the Lascaux cave. In 1998 they were introduced into the Chernobyl exclusion zone to foster biodiversity.

Sarcophagus of Chernobyl Unit 4, 27 July 2005.

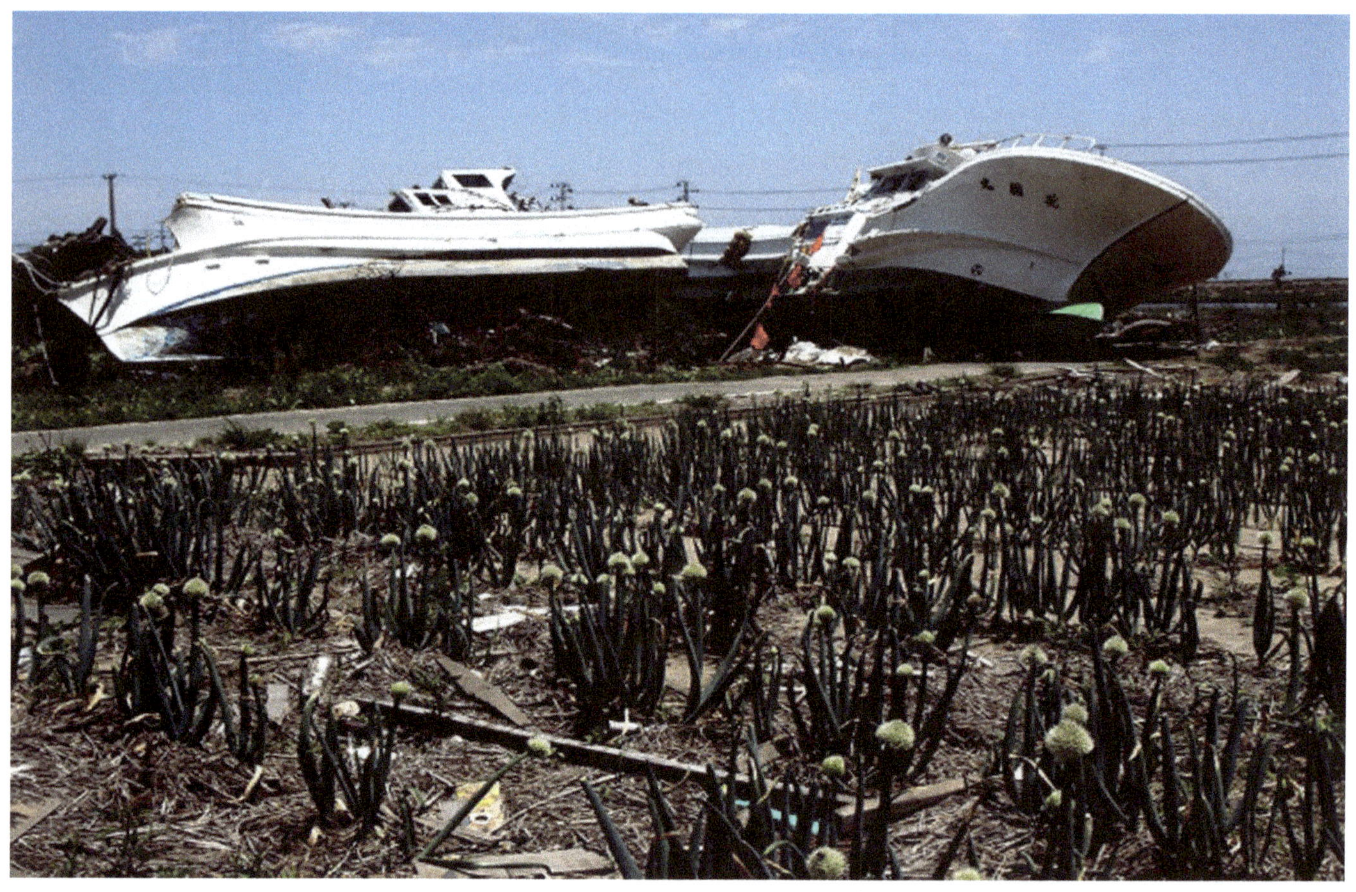

Tokyo nuclear photographer Kenji Higuchi visited Fukushima one month after the triple meltdown of the Daiichi reactors. He photographed this tsunami-tossed boat 25 km. from Ground Zero. As strange as the pairing of boats and onions appears, far more incongruous is the mix of unseen fission products with rich farmland.

Ebi Agricultural Area Kashima Kita, Minamisoma City, 7 June 2011.

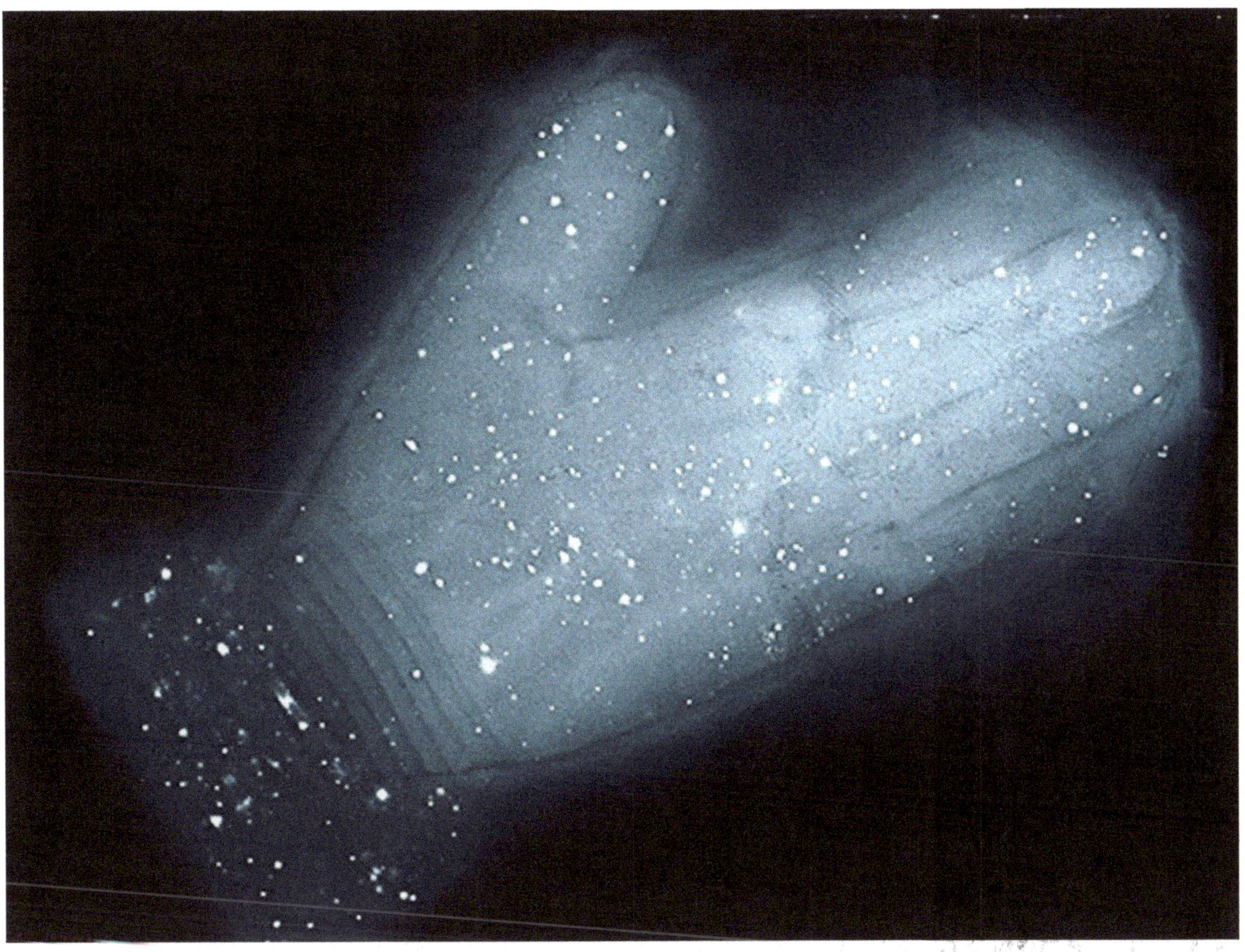

Tokyo photographer Takashi Morizumi looks for ways to make radiation visible. This image was made when he placed a Fukushima farmer's glove on top of a sheet of film. Hot particles in the glove created pinpoint exposures in the film. Morizumi believes that the ongoing radioactive menace of Fukushima is far from over. He began as a photo student of Kenji Higuchi's and has gone on to document the aftermath of Chernobyl in the Ukraine, the former Soviet nuclear test site in Kazakhstan, land impacted by a plutonium reactor in Chelyabinsk, and the medical impact of depleted uranium munitions in Iraq.

The glove belongs to Masatsugu Shiga, a cattle farmer who lived in Iitate village about 30 km. from the Fukushima Daiichi Nuclear Reactors. Mr. Shiga was using the glove while working with his cows. After the accident, Mr. Shiga moved to Minami Soma. On his farm there are no more cows, and his home and buildings are falling into ruin.

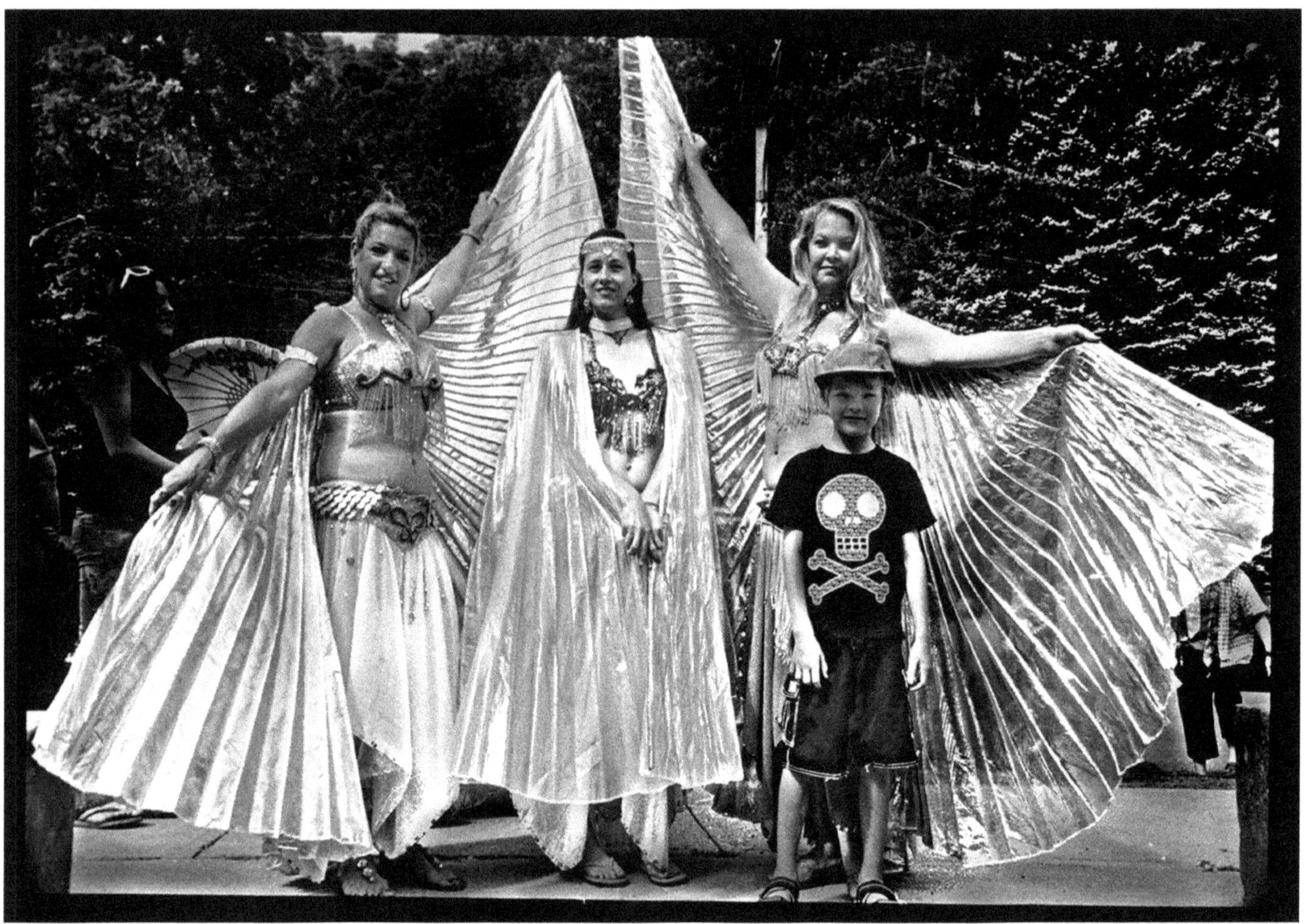

Denial. Young mothers prepare for the Canada Day Parade in the town of Port Hope, Ontario—home to the oldest and largest uranium refinery in the world. From the 1940s to the 1960s it processed uranium for the US Bomb program; since then it has processed uranium for reactors worldwide. The Cameco Corporation runs the plant and is the area's biggest employer; it bankrolls many of the town's cultural events. The 16,000 citizens of Port Hope have lived for decades under a fine mist of radioactive particles routinely released from the plant's stacks. Radiation-induced cancers and related illnesses have long latencies. Most citizens believe their town is safe, though Port Hope has, since 2013, been engaged in The Port Hope Area Initiative, a $1.28 billion cleanup of radioactive wastes throughout the area—the costliest radioactive cleanup in Canadian history.

Canada Day, Port Hope, 1 July 2012. Photo: Robert Del Tredici.

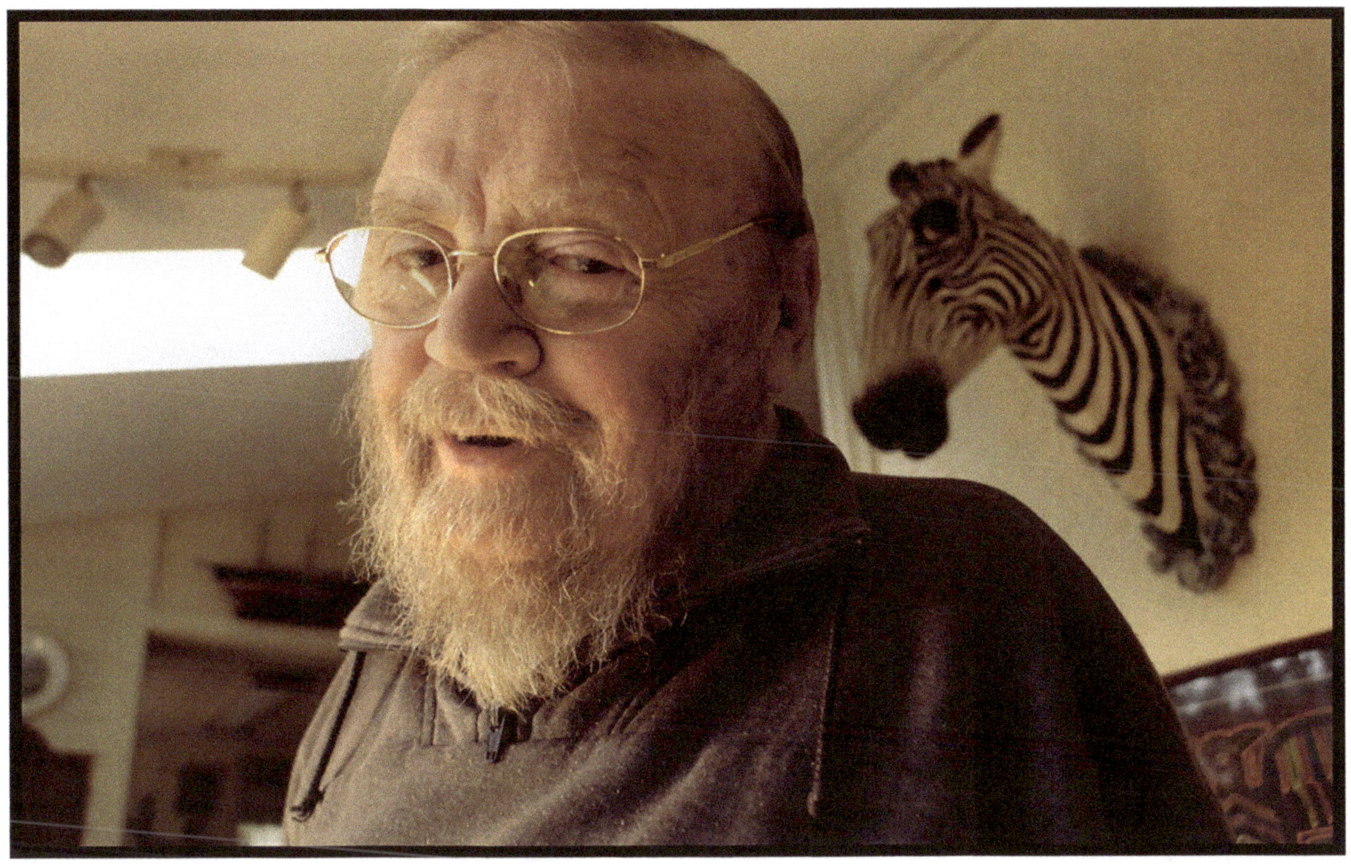

Despair. Canadian wilderness writer Farley Mowat was a long-time resident of Port Hope. Asked how he felt about the town's uranium facility, he said it was for him proof of the folly of the human species, now well beyond salvation. He favored protecting the planet's wildlife, since many species now need all the help they can get.

Farley Mowat's living room, 18 King Street, Port Hope, Ontario. 10 November 2010. Photo: Robert Del Tredici.

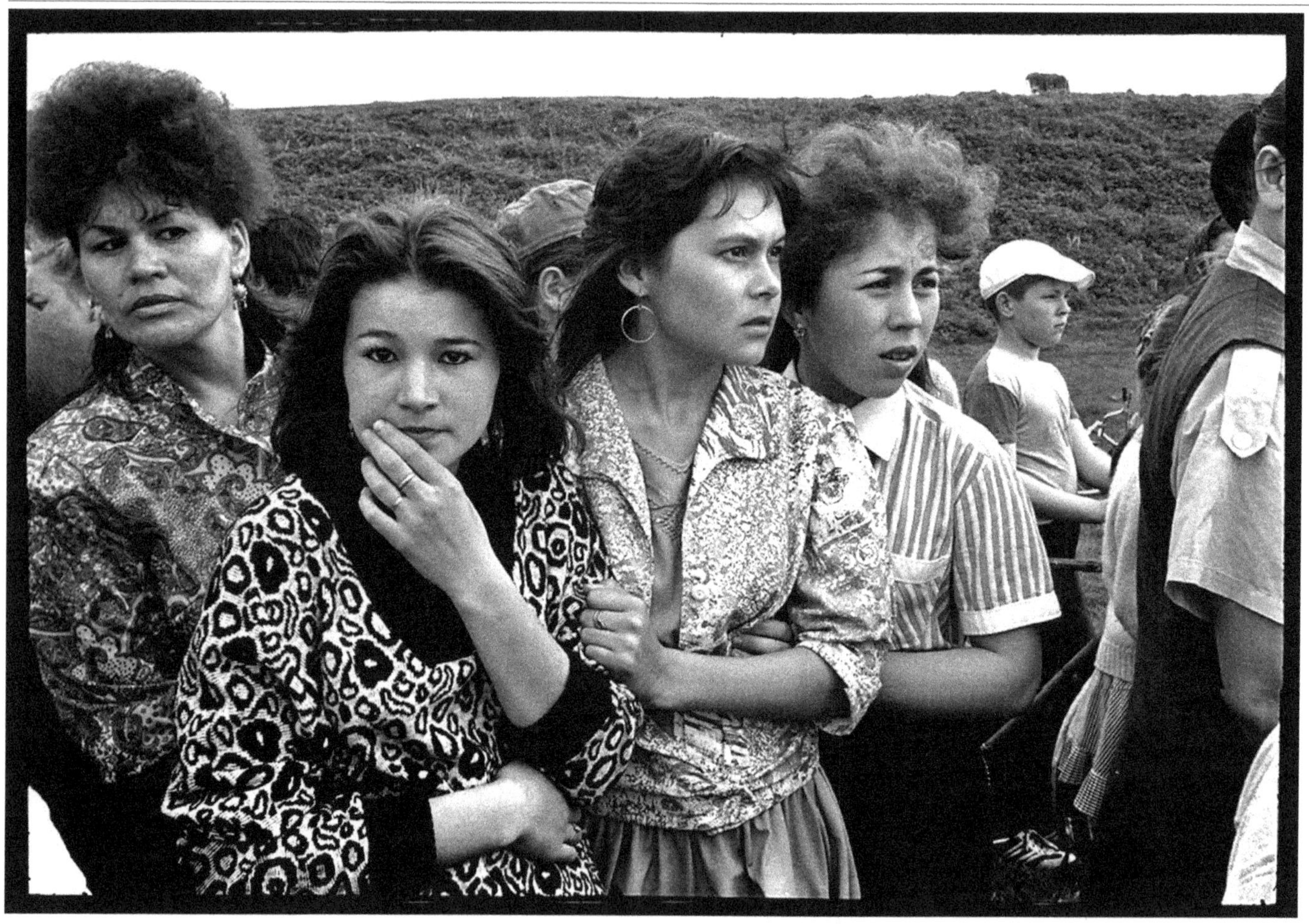

Acknowledging the hidden legacy. Maids of Muslyuomovo watch westerners measure radioactivity in the river that flows past their town. The Chelyabinsk reactor, 35 km. upstream, made plutonium for the first Soviet a-bombs. In the mid-1940's, for four years, it dumped liquid high-level radioactive waste in the Techa River. Its waters turned black; many fell ill and died. For the first time in 40 years these villagers are learning the reality behind the health crisis in their region. *Tartar-Bakshir village of Muslyuomovo on the River Techa, Chelyabinsk, Russia. 17 March 1991. Photo: Robert Del Tredici.*

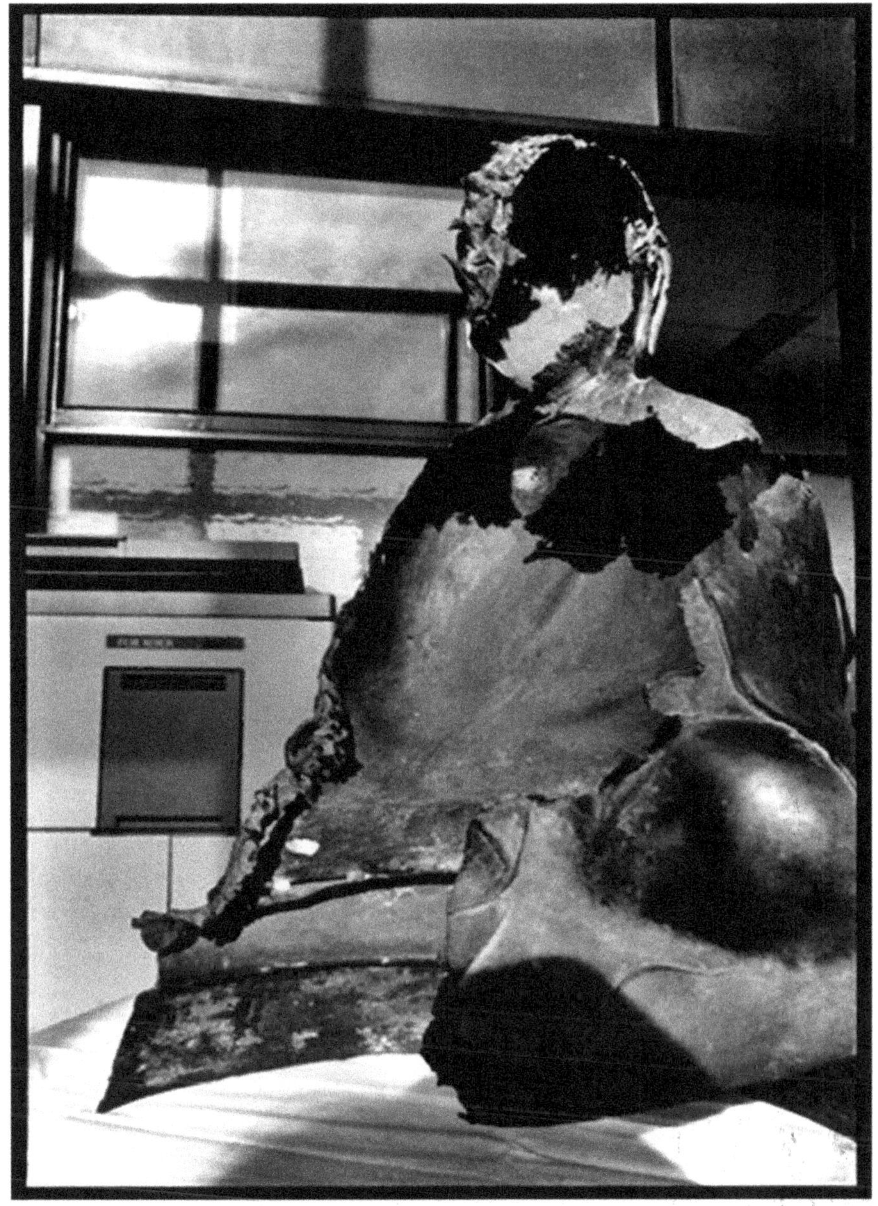

Hiroshima Buddha melted by heat from the atomic bomb. A Russian poet says of this statue that it tells us "even the gods are vulnerable to nuclear weapons." Hiroshima survivors tell us "nuclear weapons and human beings cannot coexist." Yet coexist we do; but our co-dependence on it erodes our Buddha nature and our basic humanity.

The drama of the Bomb can end in only one of two ways: everyone who wants one will get one, and we end up using it on each other; or we summon the will to abolish it. I see no middle way.

Stressed Aluminum Dome showing impact from a 1957 atmospheric nuclear blast code-named "Priscilla." Priscilla's balloon-borne Mark 15.39 "primary" device unleashed 37 kilotons of explosive power in Area 5 of the Nevada Test Site (NTS) on June 24th, 1957.

Priscilla was part of the 1957 "Plumbbob" series of 29 nuclear test explosions with 28 fission blasts and 1 two-stage thermonuclear blast (code-named "Hood" which, at 75 kilotons, was the largest atmospheric test ever at NTS). The Plumbbob series also included special exercises code-named "Desert Rock VII" and "Desert Rock VIII," which intentionally exposed 18,000 Air Force, Army, Navy, and Marine troops and 1200 pigs to nuclear blast effects.

The Plumbbob series occurred in 1957, between May 28th and October 7. Its tests in chronological order, were code-named: Boltzmann, Franklin, Lassen, Wilson, Priscilla, Coulombe-A, Hood, Diablo, John, Kepler, Owens, Pascal-A, Stokes, Saturn, Shasta, Doppler, Pascal-B, Franklin Prime, Smokey, Galileo, Wheeler, Coulomb-B, Laplace, Fizeau, Newton, Rainier, Whitney, Charlston, and Morgan.

Modes of delivery for the nuclear devices: 13 by balloon, 9 by tower, 3 in underground shafts, 2 in surface tests, 2 in tunnels (including "Rainier," the first fully contained deep underground test), and 1 ("John") mounted on a high altitude, air-to-air rocket.

According to the National Cancer Institute, "the Plumbbob tests released some 58,300 kilocuries of radioiodine (1-131) into the atmosphere, more than twice as much as any other continental test series. This produced total civilian radiation exposures amounting to 120 million person-rads of thyroid tissue exposure (about 32% of all exposure due to continental nuclear tests). This can be expected to eventually cause about 38,000 cases of thyroid cancer, leading to some 1900 deaths. National Cancer Institute Study of Estimating Thyroid doses of 1-131 Received by Americans From Nevada Atmospheric Nuclear Bomb Tests, 1997.

(from Wikipedia. *http://nuclearweaponarchive.org/Usa/Tests/Plumbbob.html*)

Shot Priscilla, 1957. Nevada Test Site. Photograph in Formed Tin and Lag Bolts Frame, James Crnkovich, 1994. From the collection of Bob Mielke.

James Crnkovich

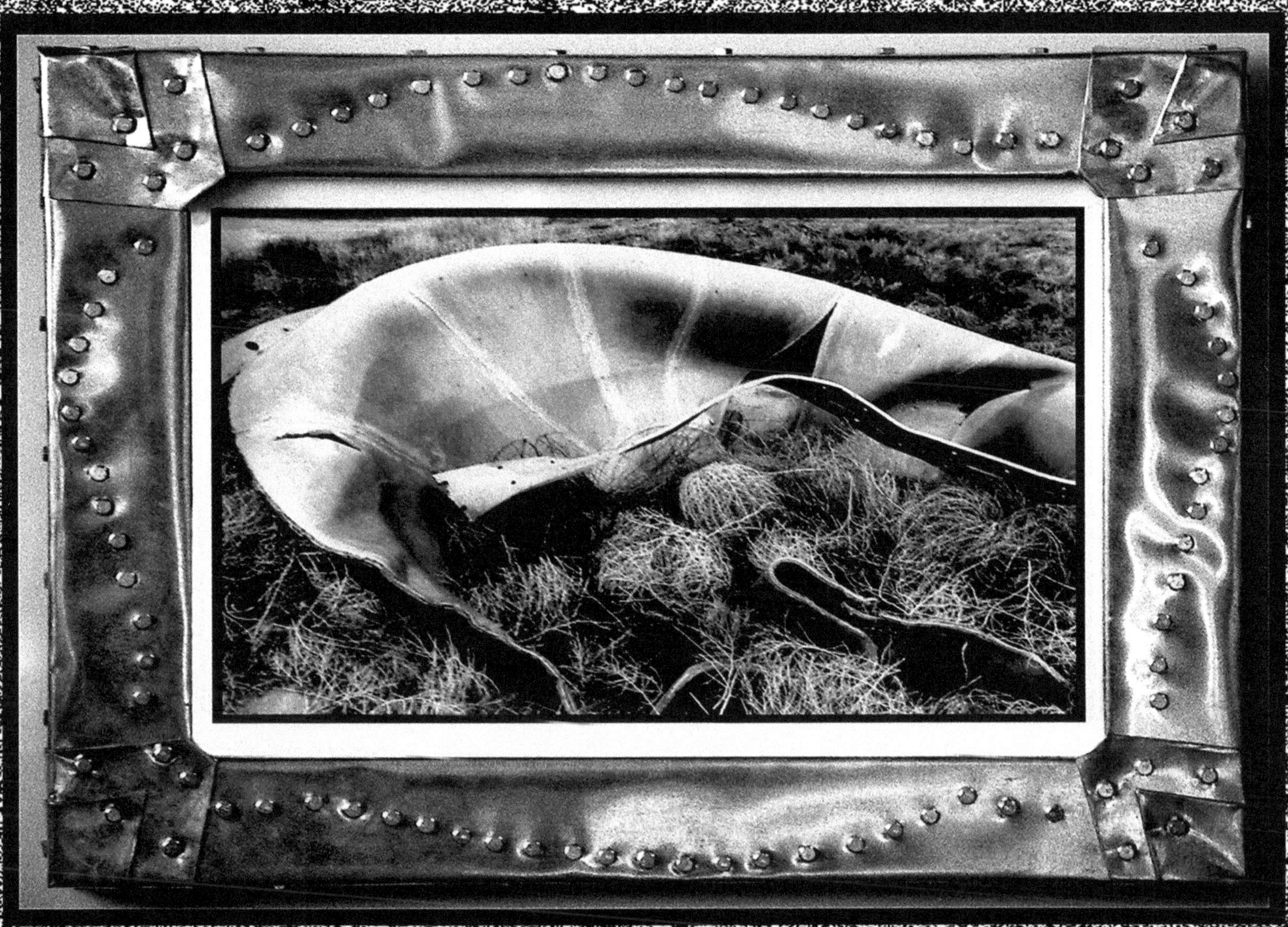

www.ingramcontent.com/pod-product-compliance
Lightning Source LLC
LaVergne TN
LVHW070935160826
845679LV00021B/1810

* 9 7 8 1 9 3 6 1 3 5 0 8 0 *